POCKET GUIDE

ROCKS & MINERALS
OF SOUTHERN AFRICA

BRUCE CAIRNCROSS

Published by Struik Nature
(an imprint of Random House Struik (Pty) Ltd)
Wembley Square, First Floor,
Solan Road, Gardens, Cape Town, 8001
PO Box 1144, Cape Town, 8000 South Africa
Company Reg. No. 1966/003153/07

Visit us at **www.randomstruik.co.za** and
subscribe to our newsletter for monthly updates and news.

First published in 2010
3 5 7 9 10 8 6 4 2

Publishing manager: Pippa Parker
Managing editor: Helen de Villiers
Editor: Emily Bowles
Design director: Janice Evans
Designer: Louise Topping
Cartographer: David du Plessis
Proofreader: Tessa Kennedy

Reproduction by Hirt & Carter Cape (Pty) Ltd
Printed and bound by Craft Print International Ltd

ISBN 978 1 77007 443 9

Front cover: Wulfenite crystals, 2.3 cm. Tsumeb mine, Namibia. **Back cover:** Mimetite crystals, 2.3 cm. Tsumeb mine, Namibia. **Title page:** Iridescent goethite, 9.5 cm. Vergenoeg mine, South Africa. **This page:** Brucite, 7 cm. Palabora mine, South Africa. **Contents page:** Quartz crystals, 5.5 cm. Kopanang mine, South Africa.

CONTENTS

INTRODUCTION

Southern Africa is endowed with a wealth of mineral deposits, as well as several internationally famous geological formations and world heritage sites. The region's geological history spans an enormous time period, with some of its rocks being over 3 000 million years old. All of the three important rock types – igneous, metamorphic and sedimentary – occur here. They are composed of many different mineral species, ranging from those that are somewhat dull to those with chemical or aesthetic properties that make them useful in industry, in building or as gems or collector specimens.

Currently over 4 400 mineral species are known worldwide, many of which are found in southern Africa. This book features the more common, interesting or important rock and mineral types and is intended to provide a quick and easy field reference for amateur geologists and collectors.

If you wish to learn more about the distribution or characteristics of rare minerals, the bibliography lists easily accessible professional publications dealing in greater detail with the geology of southern Africa.

Some of the best aquamarine and schorl tourmaline in the world come from the Erongo mountains of Namibia.

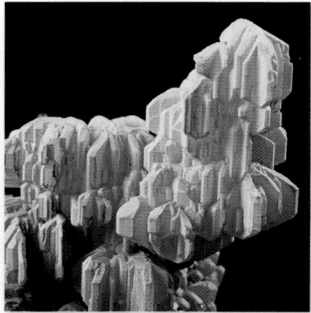

This 7.4 cm specimen of cerussite and hydrocerussite comes from the Tsumeb mine in Namibia – the source of many well-known minerals.

Minerals

A mineral is a naturally occurring inorganic solid, consisting either of a single element or a mixture of elements. All minerals have formal scientific names that have to be approved by the International Mineralogical Association (IMA), which formulates rules and guidelines for naming minerals. However, there are also unofficial common, informal names used by the trade or by amateurs, e.g. 'Cape ruby' for 'pyrope garnet'. This applies to both minerals and rocks. It is best to use the scientific names wherever possible to limit confusion.

The Chapman's Peak road (Cape Peninsula) runs along the contact between granite (below) and sedimentary rocks (above).

Weathered dolerite at the Valley of Desolation, Graaff-Reinet district, South Africa.

GEMSTONES

Gemstones are a special category of minerals and are defined by three criteria: beauty, durability and rarity. With the exception of pearls and coral, which are organic gems, all other gemstones are naturally occurring minerals or rocks.

The terms 'precious' and 'semiprecious' have historically been used to describe gems. Precious traditionally refers to diamond, ruby, emerald, sapphire and pearl, and sometimes opal and alexandrite. Semiprecious refers to all other gems, e.g. aquamarine, tiger's eye, topaz, tourmaline, amethyst and stones that are not as rare or expensive as the precious group.

However, this subdivision is discouraged and the term gemstones is now more acceptable for both categories. Gemstones are hosted in all three major types of rock and most gemstones are relatively rare. Knowing which gemstones are associated with which rock type helps prospectors to discover gemstone deposits. Nowadays, synthetic gemstones grown in a laboratory are becoming increasingly common in the market. Often they are equal in quality to the natural gemstones, but cheaper.

Traditionally, the following gemstones are associated with the months of the year:

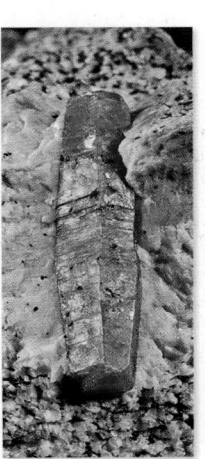

Ruby corundum crystal, 3.5 cm, in granite, a rock type that often hosts gemstones. Limpopo province, South Africa.

Birthstones			
January	Garnet	July	Ruby
February	Amethyst	August	Peridot
March	Aquamarine	September	Sapphire
April	Diamond	October	Opal
May	Emerald	November	Topaz
June	Pearl	December	Turquoise

Rocks

Top: Close-up of granite, 8 cm. Namibia.
Above: Coarse-grained granitic pegmatite, 45 cm. Klein Spitzkoppe, Namibia.

A rock is composed of a combination of minerals. Some rocks can have many mineral components, while others may consist of only a few minerals. To distinguish between different types of rocks, first the minerals that comprise them have to be studied and identified. Fortunately, the suite of rock-forming minerals is relatively small, so this identification process is not too arduous. As already mentioned, all rocks can be classified into three major groups – igneous, sedimentary and metamorphic. These three categories are described in detail in the section of this book dealing with southern African rocks (see page 122).

Identifying minerals & rocks

It requires patience, practice and wide reading to become expert at identifying rocks and minerals. While many of the common species display distinctive forms, others occur in a confusing variety of habits and colours. Quartz, for instance, is probably the most common mineral on Earth, yet it can prove frustrating to distinguish from other similar-looking species.

Aquamarine and muscovite, 8.8 cm. Erongo mountains, Namibia.

However, certain fundamental physical properties are listed in this book as a quantitative way of helping to identify mineral samples. These properties include colour, streak, lustre, transparency and translucency, form or habit, cleavage, tenacity, specific gravity and hardness, each of which is explained below. In the case of rocks, characteristics such as the size of the mineral grains, texture and, sometimes, colour can provide valuable clues.

Ettringite, 3.1 cm. N'Chwaning II mine, South Africa.

IDENTIFYING MINERALS

While some specimens are always difficult to identify, working systematically through the physical characteristics helps to narrow down the options. Some of these properties, such as streak, weight (related to specific gravity) and hardness, can be tested without sophisticated equipment. Bear in mind, though, that one attribute alone is not sufficient to make an accurate identification, so a combination of several physical attributes should be used.

Crystal systems

All of the known minerals on Earth can be classified into just six crystal systems – there are no more. Each of these systems is defined according to the length and angular relationships of imaginary internal crystal axes. Within each crystal system, however, there are many different forms. Below are the basic crystal systems from which all other structures originate.

cubic tetragonal orthorhombic monoclinic triclinic hexagonal

Physical properties of minerals

The terms used to describe the physical properties of minerals are listed and explained below.

Colour Some minerals can be identified on the basis of their unique colour, e.g. red ruby. But others, such as fluorite, beryl and quartz occur in several different colours, so this criterion is useful but not uniquely diagnostic. Exercise caution when using colour as a characteristic!

Common quartz is colourless or white, but is coloured here by inclusions of red hematite, 4.8 cm. Goboboseb Mountains, Namibia.

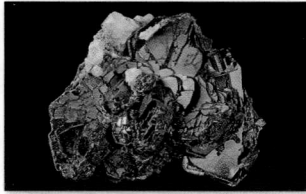

Hematite with barite and andradite garnet, 10.5 cm. Wessels mine, South Africa.

Streak When scratched on a white porcelain tile, some minerals break down, leaving behind a powdery, distinctively coloured residue known as a 'streak'. This is a useful property in certain minerals such as hematite, in which it is diagnostic. However, other minerals such as aragonite and microcline have a colourless streak.

■ **Lustre** This refers to the light reflected off the surfaces or crystal faces of a mineral. Lustre can be described as metallic, vitreous (like broken glass), resinous (like resin), pearly, silky or adamantine (the lustre of a diamond). A metallic lustre usually suggests that the mineral is a metal sulphide or oxide.

■ **Transparency and translucency (diaphaneity)** This refers to the degree to which a mineral transmits light. Some minerals and gemstones are transparent and allow one to view objects through them. Other minerals are opaque, or non-transparent. Translucent is the term for a mineral that transmits light, but through which distinct images can't be seen.

A selection of faceted southern African gemstones – tourmaline, garnet, aquamarine and heliodor.

■ **Form or shape** This describes how clearly defined the crystal faces are, and depends on criteria such as the temperature at which minerals crystallize or the chemical composition of the solution from which they have crystallized. If the mineral is well crystallized and shows clearly defined crystal faces, it is said to be euhedral. Such crystals are also called

Euhedral calcite

Subhedral chalcocite (above) and anhedral kaolinite (top).

'floaters' because they crystallized while unattached to any rock, simply floating in a liquid. If the mineral is only partially well formed, it is called subhedral. Minerals can also form without any apparent well-defined crystal faces. These are termed amorphous, anhedral or massive.

Habit This refers to particular crystalline shapes, e.g. fibrous, cauliform (shaped like a cauliflower), acicular (needle-like), bladed, tabular, prismatic, stellate (crystals radiating out from a central point), botryoidal (resembling a bunch of grapes), reniform (kidney shaped), dendritic (like a fern leaf) and arborescent (like the branches of a tree). The habit of a mineral is determined largely by its chemical composition, the arrangement of atoms and by the space available. Limited space produces distorted or 'deformed' crystals. Some crystals are even twinned, i.e. they have multiple faces and forms.

Stellate (star-like) aragonite, 9.4 cm. Gauteng, South Africa.

Bladed kyanite crystal, 6.8 cm. Zimbabwe.

Cleavage This refers to the way a mineral fractures or splits when broken. The cleavage planes or patterns are controlled by the internal atomic arrangement of elements within the mineral. Some minerals have very poor cleavage and do not break along flat planes, e.g. quartz, which displays a conchoidal fracture pattern. Other minerals, like mica, split very easily into thin sheets, along well-defined cleavage surfaces.

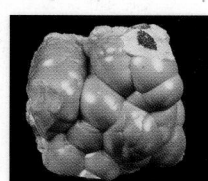

Acicular manganite crystals, 4.5 cm. N'Chwaning II mine, South Africa.

Tenacity This refers to how a mineral reacts to being crushed. For instance, it may be brittle and shatter, or tenacious and resist breaking, malleable like gold or flexible like asbestos.

Specific gravity (SG) In general, the SG of a mineral relates to the atomic weight of the constituent elements, and how they are packed together. The SG is the ratio of the weight of a stone to that of an equal volume of water. Some examples

Botryoidal aragonite 'dripstone', 6 cm. Montrose mine, South Africa.

Barite (left) has a higher specific gravity than calcite (right), although they can look very similar.

of SG are gold, 19.3, quartz, 2.66, and topaz, 3.5. Knowing the SG of minerals can be key in helping to identify them. The greater the SG, the heavier the species, e.g. barite and calcite may superficially resemble each other, but barite has a much greater SG and is heavier.

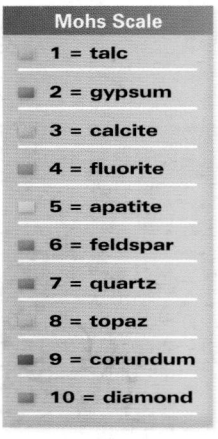

Mohs Scale
1 = talc
2 = gypsum
3 = calcite
4 = fluorite
5 = apatite
6 = feldspar
7 = quartz
8 = topaz
9 = corundum
10 = diamond

Hardness The hardness of a mineral can be a useful clue to its identity. You can compare the hardness of a given specimen with the minerals defined in Mohs Scale, which grades 10 minerals from the softest, talc (1), to the hardest, diamond (10).

Corundum, for instance, can scratch all other minerals except diamond. Fluorite can scratch all minerals softer than its hardness value (4). In practice, some common objects can be used to test the hardness of minerals, e.g. a fingernail (hardness about 2.5) can scratch talc and gypsum. Common window glass has a hardness of about 5.5–6. An ordinary penknife blade has a hardness of about 6. Most gemstones, by definition, are relatively hard.

Physical properties of minerals: summary table	
Property	**Description**
Colour	Varied; some species multicoloured, others not
Streak	Trace remains when scratching mineral on a white tile
Lustre	The way in which light is reflected off the mineral surface
Transparency/ translucency	The degree to which a mineral transmits light
Form / shape	The outer appearance of the mineral
Habit	The specific shapes of minerals (see text)
Cleavage	The way in which a mineral splits or breaks when broken
Tenacity	How a mineral reacts when crushed
Specific gravity (SG)	Related to the atomic structure of the mineral – the greater the SG, the heavier the mineral
Hardness	Mohs Scale of hardness ranges from 1 to 10 (softest to hardest)

IDENTIFYING ROCKS

Identifying rocks begins with determining how they formed, i.e. whether they are igneous, sedimentary or metamorphic. Clues may include their location, grain size, hardness, texture and colour. One also needs to study and recognize the minerals that comprise them.

Wind-blown sand dunes in the Namib Desert, en route to Sossusvlei, Namibia.

How to use this book

The book is divided into two parts: minerals (pages 12–121) and rocks (pages 122–153).

Mineral species are listed in alphabetical order. Each account includes a tinted summary table listing the physical characteristics of the mineral and, where relevant, **properties that are diagnostic are highlighted by a darker band of colour**.

The following icons depict important additional information:

Key to icons		
Descriptor	**Icon**	
Rock-forming mineral		Found as a component in rocks
Collector species		Attractive and desirable to collectors
Gemstone		Transparent and hard enough to cut and polish
Industrial/economic use		Of economic importance; often mined
Occurrence bar		Shaded from rare (one block) to most common (four blocks)

Each species account provides text on the description, uses (if any) and occurrence of the mineral. The following abbreviations have been used for the names of provinces within South Africa:

NC: Northern Cape; LIM: Limpopo; MP: Mpumalanga; GAU: Gauteng; NW: North West; KZN: KwaZulu-Natal; FS: Free State; WC: Western Cape; EC: Eastern Cape.

The rocks are organized by rock type: igneous (page 124), sedimentary (page 134) and metamorphic (page 146). Accounts are in alphabetical order within each of these sections. The text for each species gives details about the colour, composition and occurrence of each rock.

MINERALS

In this book, we've grouped the minerals alphabetically. However, chemical composition is the usual way in which mineralogists classify minerals. All minerals are made up of elements (see the alphabetical list opposite): either a single element, as is the case with gold, or, more usually, a combination of elements. Each mineral has both positive and negative ions – called cations and anions respectively. Mineralogists group minerals according to the anion classification system, based on those anions they have in common.

Of more than 60, here are the most common groups, in order of their general abundance in the Earth's crust:

- **silicates (most abundant)** – all contain the (Si) and (O) anions
- **carbonates** – all contain the $(CO_3)^{2-}$ anion
- **sulphates** – all contain the $(SO_4)^{2-}$ anion
- **halides** – all contain at least one halogen anion, F^-, Cl^-, Br^-, I^-.
- **oxides** – all contain an O^{2-} anion
- **sulphides** – all contain an S^{2-} anion
- **phosphates** – all contain the $(PO_4)^{3-}$ anion

The chemical formula given for any mineral expresses the elements that are present in it, and in what proportions. Subscripts give the number of atoms of a given element in the mineral, while the superscripts indicate the charge of the ion (**+** for a positive charge, **–** for a negative one).

So, $CaCO_3$ (the formula for the carbonate mineral aragonite) shows that this mineral consists of one calcium atom, one carbon atom and three oxygen atoms. (Note that when the CO_3 carbonate anion (negative) combines with the calcium cation (positive), the charge becomes neutral, which is why no superscript number appears.)

A 65-carat yellow octahedral diamond, a 9-carat pink diamond and some smaller diamonds. Barkly West, South Africa.

Alphabetical list of chemical elements

Symbol	Name	Symbol	Name	Symbol	Name
Ac	Actinium	H	Hydrogen	Pt	Platinum
Ag	Silver	He	Helium	Pu	Plutonium
Al	Aluminium	Hf	Hafnium	Ra	Radium
Am	Americium	Hg	Mercury	Rb	Rubidium
Ar	Argon	Ho	Holmium	Re	Rhenium
As	Arsenic	Hs	Hassium	Rf	Rutherfordium
At	Astatine	I	Iodine	Rh	Rhodium
Au	Gold	In	Indium	Rn	Radon
B	Boron	Ir	Iridium	S	Sulphur
Ba	Barium	K	Potassium	Sb	Antimony
Be	Beryllium	Kr	Krypton	Sc	Scandium
Bh	Bohrium	La	Lanthanum	Se	Selenium
Bi	Bismuth	Li	Lithium	Si	Silicon
Bk	Berkelium	Lr	Lawrencium	Sg	Seaborgium
Br	Bromine	Lu	Lutetium	Sm	Samarium
C	Carbon	Lw	Lawrencium	Sn	Tin
Ca	Calcium	Md	Mendelevium	Sr	Strontium
Cd	Cadmium	Mg	Magnesium	Ta	Tantalum
Ce	Cerium	Mn	Manganese	Tb	Terbium
Cf	Californium	Mo	Molybdenum	Tc	Technetium
Cl	Chlorine	Mt	Meitnerium	Te	Tellurium
Cm	Curium	N	Nitrogen	Th	Thorium
Co	Cobalt	Na	Sodium	Ti	Titanium
Cr	Chromium	Nb	Niobium	Tl	Thallium
Cs	Cesium	Nd	Neodymium	Tu	Thulium
Cu	Copper	Ne	Neon	U	Uranium
Db	Dubnium	Ni	Nickel	Uub	Ununbium
Ds	Darmstadtnium	No	Nobelium	Uuh	Ununhexium
Dy	Dysprosium	Np	Neptunium	Uuq	Ununquadium
Er	Erbium	O	Oxygen	Uuu	Ununium
Es	Einsteinium	Os	Osmium	V	Vanadium
F	Fluorine	P	Phosphorus	W	Tungsten
Fe	Iron	Pa	Protactinium	Xe	Xenon
Fm	Fermium	Pb	Lead	Y	Yttrium
Fr	Francium	Pd	Palladium	Yb	Ytterbium
Ga	Gallium	Pm	Promethium	Zn	Zinc
Gd	Gadolinium	Po	Polonium	Zr	Zirconium
Ge	Germanium	Pr	Praseodymium		

■ Aegirine

$NaFe^{3+}Si_2O_6$

Composition	silicate
Crystal system	monoclinic
Hardness	6
Specific gravity	3.55–3.60
Streak	pale yellow-grey
Lustre	vitreous, resinous

■ **Description** This member of the pyroxene group is a highly lustrous, dark green to black mineral that forms prismatic crystals and fibrous masses. It is named after Aegir, the Scandinavian god of the sea.

■ **Occurrence** Occurs in intermediate, alkali-rich, silica-undersaturated rocks, such as syenites. Common in many of the southern African alkaline complexes. **South Africa** NW: Occurs in the Pilanesberg. **Namibia** Euhedral crystals of aegirine are found in the Aris phonolite. **Zimbabwe** Occurs at Chishanya. **Malawi** The finest crystals of aegirine in the southern African region, and perhaps globally, come from the Zomba Mountain.

Aegirine crystal, 10.7 cm.
Zomba Mountain, Malawi.

Aegirine crystals, 4.8 cm,
Zomba, Malawi.

Almandine ■

$Fe_3^{2+}Al_2Si_3O_{12}$

Composition	silicate
Crystal system	cubic
Hardness	7–7.5
Specific gravity	4.1– 4.3
Streak	white
Lustre	vitreous, resinous

■ **Description** A member of the garnet group. Usually red but can occur as black crystals. Forms a chemical series with two other garnet species, pyrope and spessartine.

■ **Uses** Used in sandpaper and cloth abrasives because of its hardness.

■ **Occurrence** Occurs in metamorphic rocks, schists and gneisses and is common in pegmatites. **South Africa** NC: Found on several farms in the Gordonia district. LIM: Almandine has been proven in the Soutpansberg district 30 km south of Musina, close to Emakhazeni. **Namibia** Widespread in schists in the districts around Karibib, Usakos and Swakopmund and in the Kuiseb schist belt. Also occurs in other metamorphic terrains. **Botswana** Found in many localities in the metamorphic rocks in eastern Botswana. **Zimbabwe** Mined as a gemstone in the Beitbridge area, near Karoi and in the northeastern parts of the country. Other noteworthy prospects are Sekuru and Manyuchi in the Mudzi district. Occurs as crystals of up to 8 cm from the Miami mine, Hurungwe district.

Dodecahedral almandine garnet crystal, 3.4 cm. Zimbabwe.

Andalusite

Al_2SiO_5

Composition	silicate
Crystal system	orthorhombic
Hardness	6.5–7.5
Specific gravity	3.13–3.16
Streak	white
Lustre	vitreous

Description A common metamorphic mineral in slates and schists, forming elongate pencil-like red-brown to tan crystals. It forms under relatively low crustal pressure and temperature. A variety of andalusite, chiastolite, contains carbonaceous inclusions in the form of an 'X' that can be seen when the mineral is cut in cross-section.

Uses It is used in the manufacture of refractory products and in the ceramics industry. Aluminium – a lightweight metal used in the aircraft and aerospace industry – is extracted from it. Aluminium also has applications as an anticorrosive agent.

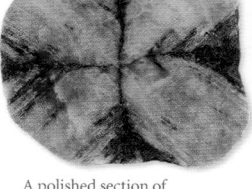

A polished section of andalusite, variety chiastolite, 4 cm. Zimbabwe.

Occurrence **South Africa** NC: Found in pegmatites near Kakamas, Mottelsrivier and Bokvasmaak. LIM/MP: Large deposits occur around Thabazimbi, Chuniespoort, Penge and Lydenburg. NW: Good deposits also found at Groot Marico and Zeerust in the Marico district. FS: Andalusite occurs in the metamorphic rocks surrounding Parys and Vredefort. **Namibia** Common in schists southwest of Okahandja, and in the Kuiseb River Valley. **Botswana** Found in the eastern districts. **Zimbabwe** Schist and gneiss north of Mtoko and in the Mwami and Masvingo districts contain andalusite. Chiastolite crystals occur in the Hurungwe and Mwami districts. **Swaziland** Deposits found in pyrophyllite schists near Sicunusa (which were mined as a source of aluminium).

Andalusite crystals in hornfels, 32 cm. Penge, South Africa.

Andradite

$Ca_3Fe_2^{3+}(SiO_4)_3$

Composition	silicate
Crystal system	cubic
Hardness	6.5–7
Specific gravity	3.7–4.1
Streak	white
Lustre	vitreous, resinous

Description A member of the garnet group. Usually found as dark red dodecahedral crystals, similar to almandine garnet. Forms a chemical series with three other garnet species: grossular, morimotoite and schorlomite. Demantoid is the green chrome-rich gem variety of andradite.

Uses Demantoid is used as a gemstone.

Occurrence **South Africa** NC: Brown andradite crystals occur in Namaqualand, at Doringkraal. LIM: Crystals have been found in the Soutpansberg. NW: Green andradite occurs in the Kalahari manganese field near Rustenburg. **Namibia** Brown to black andradite is found in the Otjosondu manganese deposits northeast of Okahandja. It is also found in some pegmatites. Demantoid garnet occurs in marble west of Erongo, central Namibia. **Zimbabwe** Dark brown to black andradite is found in the Chimanda Communal Lands near the Mazowe River, Bulawayo district.

Andradite garnet, variety demantoid, 2 cm. Tubussis, Namibia.

Anglesite

PbSO$_4$

Composition	sulphate
Crystal system	orthorhombic
Hardness	2.5–3
Specific gravity	6.38
Streak	white
Lustre	adamantine, glassy

Description Typically colourless to white. It forms heavy tabular or prismatic crystals or nodular stalactitic masses. Found as a secondary mineral associated with lead deposits. Minor elemental impurities colour the mineral – green (copper), yellow (cadmium) or pink (cobalt).

Occurrence **South Africa** Rare. NC: Some anglesite was collected at the Aggeneys mine. GAU: Small crystals collected from the Argent mine, near Delmas. Other isolated deposits occur in the province. NW: Small crystals occur in scattered lead-zinc deposits in dolomite host rock, e.g. in the Marico district. **Namibia** Beautiful, coloured crystals came from the Tsumeb mine. **Zimbabwe** Rare – occurs in the Mutare, Belingwe and Bulawayo districts.

Large cluster of anglesite crystals, 15 cm. Tsumeb mine, Namibia.

Anglesite crystal, 1.2 cm. Tsumeb mine, Namibia.

Antimony
Sb

Composition	native element
Crystal system	hexagonal
Hardness	3–3.5
Specific gravity	6.7
Streak	grey
Lustre	metallic

Description A bright silver or tin-white metal.

Uses Antimony has a fairly low melting point (631 °C). Alloyed with lead, it is used in batteries. It is also used in ammunition, solder and bearings. A form of antimony may be used as a flame-retardant in textiles, plastics and rubber.

Occurrence South Africa LIM: Crystals of antimony have been found in the Murchison greenstone belt at Gravelotte. MP: Small amounts were found at Barberton and Malelane. **Zimbabwe** Antimony has been reported from the Kwekwe area.

Native antimony, 4.6 cm. Consolidated Murchison mine, South Africa.

Native antimony, 6.1 cm. Consolidated Murchison mine, South Africa.

Apatite-(CaF)

$Ca_5(PO_4)_3F$

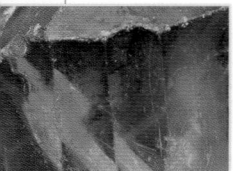

Composition	phosphate
Crystal system	hexagonal
Hardness	5
Specific gravity	3.1–3.2
Streak	white
Lustre	vitreous, resinous

Description The crystals are normally six-sided, prismatic and can be elongate or short and stubby. Apatite commonly fluoresces under ultraviolet light. Its colour is variable: white to grey, yellow, green, blue, violet, purple, red or brown.

Uses The phosphate produced from it is used in mineral feedstock, detergents, pharmaceutical products and fertilizers.

Occurrence This common and widespread mineral occurs in many igneous and metamorphic rocks and in hydrothermal vein deposits, alkaline complexes, carbonatites and pegmatites. **South Africa** NC: Found in pegmatites, e.g. at Straussheim and Blesberg. LIM: Apatite-(CaF) comes from the Palabora mine, the Soutpansberg and pegmatites in the Letaba district. NW: Occurs in the Pilanesberg. KZN: It is reported from Marble Delta. **Namibia** Found in many pegmatites and carbonatites. **Botswana** Occurs east of Mmadinare. **Zimbabwe** Widespread – relatively common in granites, pegmatites and alkaline rocks. **Swaziland** Some pegmatites here contain the mineral.

Apatite-(CaF), 2.4 cm. Zimbabwe.

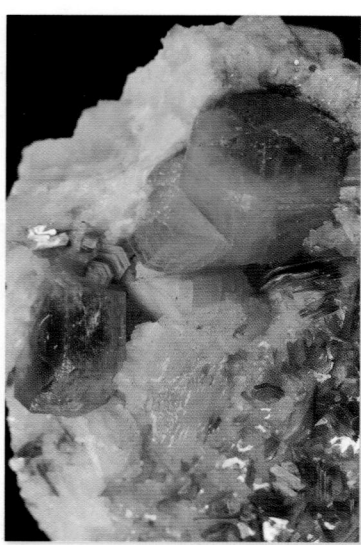

Apatite-(CaF) crystals with muscovite, 3.2 cm. Blesberg mine, South Africa.

Apophyllite-(KF)

$KCa_4Si_8O_{20}F \cdot 8H_2O$

Composition	silicate
Crystal system	tetragonal
Hardness	4.5–5
Specific gravity	2.37
Streak	white
Lustre	pearly, vitreous

Description The crystals have a variety of shapes: pseudocubic, tabular, pyramidal or elongate prisms, and are usually white, cream or colourless. Apophyllite-(KF) is often associated with zeolite minerals.

Occurrence **South Africa** NC: Colourless apophyllite-(KF) comes from the Kimberley, Bultfontein, Jagersfontein and De Beers mines. LIM: Well-formed crystals were found at the Palabora mine (with prehnite, calcite and mesolite). GAU: Rare – some specimens have been collected at the Premier diamond mine and at the Witwatersrand gold mines. **Namibia** Found in amygdales (cavities) in the Aris phonolites south of Windhoek. **Botswana** Rare. Beautiful tabular crystals were found in a fissure at the Selebi-Phikwe mine. **Zimbabwe** Occurs as well-formed crystals at the Ayrshire mine, the Jeppe copper mine, and the Beatrice mine, and in basalts in the Victoria Falls area.

Apophyllite-(KF), 4 cm. Mponeng mine, South Africa.

Apophyllite-(KOH)

$KCa_4Si_8O_{20}(OH,F) \cdot 8H_2O$

Composition	silicate
Crystal system	tetragonal
Hardness	4.5–5
Specific gravity	2.37
Streak	white
Lustre	pearly, vitreous

Description Forms distinctive tabular, pseudocubic, colourless to white crystals, with a characteristic pearly lustre. The crystals tend to cleave easily and the corners are often bevelled.

Occurrence **South Africa** NC: Relatively common pale pink, cream, tan and colourless crystals (some reaching up to 15 cm) occur in the Kalahari manganese field mines, notably the Wessels and N'Chwaning I and II mines. MP: Flat, tabular, transparent, colourless crystals are found in faults in the Bushveld Complex gabbro, Belfast Granite Quarry (which is mined for dimension stone).

Apophyllite-(KOH), 15 cm. Wesselton mine, Kimberley, South Africa.

Apophyllite-(KOH), 5.6 cm. N'Chwaning II mine, South Africa.

Aragonite

CaCO$_3$

Composition	carbonate
Crystal system	orthorhombic
Hardness	3.5–4
Specific gravity	2.95
Streak	white
Lustre	vitreous, resinous

Description Occurs as colourless, transparent or white acicular crystals, but may also be pyramidal or tabular. Aragonite crystals are commonly found in caves as stalactites, stalagmites and dripstones.

Occurrence **South Africa** Found in many carbonate-rich limestone and dolomite caves. LIM/MP/GAU: Beautiful specimens occur at Echo Caves, Makapansgat, Sudwala Caves and Sterkfontein Cave. Found rarely in the Witwatersrand gold mines. WC: Aragonite found in the Cango Caves. **Namibia** Crystals occur at the Tsumeb and Kombat mines. Three chemical variations of aragonite are found at Tsumeb – green cupro-aragonite, white lead-rich tarnowitzite and yellow nicholsonite, which contains zinc. It is present in caves and cavities in the dolomites of the Otavi mountainland. Lenses of solid aragonite are found in the Swakopmund and Karibib districts, near Rössing. Aragonite has been worked north of Swakopmund. **Zimbabwe** Found at Mangula and the Ethel mine, Lomagundi district, as well as at the Chinoyi Caves.

Aragonite, varieties nicholsonite (left, 5.1 cm) and tarnowitzite (right). Tsumeb mine, Namibia.

Sprays of white aragonite on a brown aragonite crystal matrix, 9.6 cm. Gauteng, South Africa.

Asbestos – see grunerite and riebeckite.

Azurite

$Cu_3(CO_3)_2(OH)_2$

Composition	carbonate
Crystal system	monoclinic
Hardness	3.5–4
Specific gravity	3.77
Streak	blue
Lustre	vitreous

Description Vivid blue, varying from shades of pale blue to blue-black. Forms tabular crystals. Azurite is a common secondary mineral associated with copper deposits. It can be chemically unstable and converts to malachite by pseudomorphism, the process whereby one mineral chemically replaces another, molecule for molecule, but retains the original mineral's crystal shape. It is not a spontaneous reaction, but takes place over a long period of time.

Occurrence **South Africa** NC: Occurs as an accessory mineral in the copper mines in the Springbok/Okiep copper district. LIM: Found at the Messina mines. MP: Present in the Pilgrim's Rest region. GAU: Occurs in copper deposits and at the Willows and Vergenoeg mines. **Namibia** Localities include the Tsumeb and Tschudi mines, Otavi mountainland, the Onganja copper mine, Windhoek district, and other Namibian copper deposits. **Botswana** Rarely found. **Zimbabwe** Azurite is associated with many copper deposits throughout the country. **Swaziland** Occurs east of Nkambeni.

Azurite nodule, 5.2 cm. Zambia.

Relatively dull blue, interlocking crystals of azurite, 3 cm. Tsumeb mine, Namibia.

Highly lustrous azurite crystals, 8.7 cm. Tsumeb mine, Namibia.

Barite

BaSO$_4$

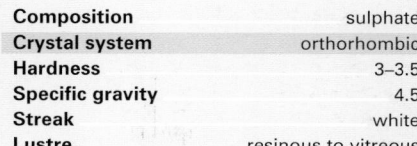
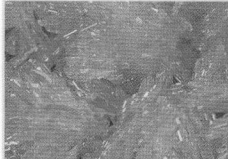

Composition	sulphate
Crystal system	orthorhombic
Hardness	3–3.5
Specific gravity	4.5
Streak	white
Lustre	resinous to vitreous

Description Barite is heavy and commonly forms prismatic tabular colourless crystals. It can also form fibrous masses or granular aggregates that are colourless, white, yellow or orange.

Uses Barite is the main source of barium – used in paint, paper, plastics, rubber, glass and ceramics. It also has medicinal uses. It is used extensively as an oil-drilling mud, i.e. the barium increases the density of the mud and lubricates the drill bits when drilling through rock.

Occurrence Usually occurs in hydrothermal veins in sedimentary and/or igneous rocks. **South Africa** NC: Occurs in the Gamsberg. Beautiful crystals come from the Kalahari manganese fields. The Okiep copper district produced orange-yellow crystals. **Namibia** Found in southern, central and northeast Namibia.
Botswana Barite occurs northeast of the central district and at Suping, 10 km north of Molepolole.
Zimbabwe Deposits occur in the Shamva district and at Bumburudza, Mwenezi district. **Swaziland** Has been mined at Mbabane district, south of Oshoek and is present in the Piggs Peak region.

Spear-shaped barite crystals, 4.2 cm. Gross Brukkaros, Namibia.

Gemmy, yellow barite, 5.6 cm. Rosh Pinah mine, Namibia.

Beryl
$Be_3Al_2Si_6O_{18}$

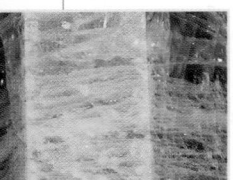

Composition	silicate
Crystal system	hexagonal
Hardness	7.5–8
Specific gravity	2.6–2.9
Streak	white
Lustre	vitreous

Description Beryl forms well-developed hexagonal crystals with flat (pinacoid) terminations. Common beryl is characteristically pale green, but can also be white to colourless. The beryl varieties are:

- aquamarine (blue-green from Fe^{2+} and Fe^{3+} ions)
- emerald (green from Cr^{3+} ions)
- heliodor (yellow from Fe^{3+} ions)
- morganite (pink from Mn^{2+} ions), and
- goshenite (colourless).

Uses Beryl is an important source of beryllium, which – alloyed with copper and aluminium – is used in the manufacture of springs, armour-piercing bullets, tools that do not produce sparks and some percussion instruments. Beryllium is also used in certain components in nuclear reactors. The coloured varieties are sold as gemstones.

Beryl variety emerald, 4.4 cm.
GEM mine, South Africa.

Beryl, variety morganite, 5.8 cm.
Mozambique.

Beryl, variety aquamarine, 2 cm.
Erongo mountains, Namibia.

◾ Occurrence

Found almost exclusively in pegmatites. **South Africa** NC: Beryl is common in pegmatites from Upington to Steinkopf. Notable deposits include the aquamarine in Namaqualand and the morganite that occurs in pegmatites at Steyns Puts. LIM: Beryl is common in pegmatites in the province: morganite is found near Leydsdorp, heliodor between Musina and Polokwane and emeralds at Gravelotte. MP: Beryl occurs in some pegmatites in the province. **Namibia** Common beryl occurs in pegmatites from Brandberg West-Uis in the north, to Sandamap-Karibib-Usakos in the south, and in Tantalite Valley. Heliodor is found close to Rössing Siding, while world-class aquamarine comes from Erongo mountains, and morganite from the Rubikon, Helikon and Neu Schwaben pegmatites. **Zimbabwe** Common beryl is widespread in northern Zimbabwe. Aquamarine and heliodor occur in the Mwami district, morganite at the Pope claims, Goromonzi district, and emeralds at Mweza (Sandawana). **Swaziland:** Beryl is found in the foothills of the Sinceni mountains.

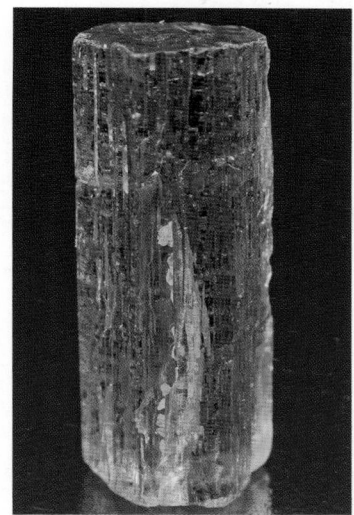

Beryl, variety heliodor, 2.8 cm. Green Walking Stick mine, Zimbabwe.

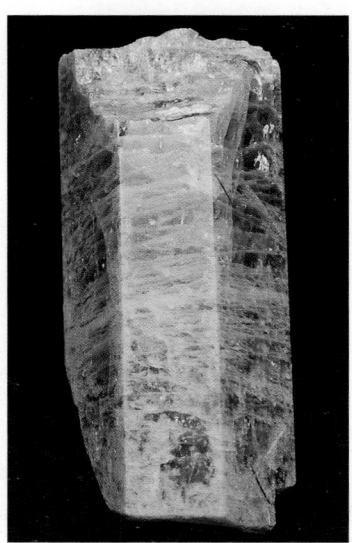

Common beryl, 7.2 cm. St Ann's mine, Zimbabwe.

Biotite

$K(Mg,Fe^{2+})_3(Al,Fe^{3+})SiO_{10}(OH,F)_2$

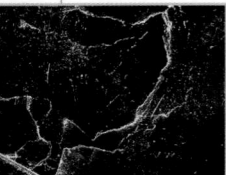

Composition	silicate
Crystal system	monoclinic
Hardness	2.5–3
Specific gravity	2.7–3.4
Streak	white
Lustre	vitreous

■ **Description** Biotite is a member of the mica group of minerals. It forms hexagonal black or dark brown to red-brown crystals that are micaceous, i.e. they are characterized by very thin paper-like layers. It is a common mineral in many igneous and metamorphic rocks.

■ **Occurrence** Biotite is found in all those southern African countries where metamorphic and igneous rocks occur and is particularly common in carbonatites in **South Africa**, **Namibia** and **Zimbabwe**. It occurs in granite, gneiss and schist, with large crystals occurring in coarse-grained pegmatites.

A hexagonal biotite crystal, 6 cm. Namibia.

Shiny iridescent biotite crystals in a metamorphic gneiss, 7 cm. Margate, South Africa.

Very large biotite crystal, 14.5 cm. Madagascar.

Bornite

Cu_5FeS_4

Composition	sulphide
Crystal system	orthorhombic
Hardness	3
Specific gravity	5.08
Streak	grey-black
Lustre	metallic

See associated minerals – chalcocite, chalcopyrite and copper.

■ **Description** Bornite gets its common name, 'peacock ore', from its distinctive colouring: a bright metallic purple mixed with orange, yellow and red. It seldom forms crystals, usually occurring as massive lumps and aggregates.

■ **Uses** It is a relatively common copper ore and therefore an important economic mineral. It is found in many copper sulphide deposits, usually associated with chalcocite and chalcopyrite.

■ **Occurrence** **South Africa** NC: It was common in the Okiep copper district. LIM: Bornite occurs in nearly all the copper mines, e.g. the Messina mines, but is less abundant than chalcopyrite, the main copper-bearing ore. **Namibia** A major copper ore, it is found at the Rosh Pinah, Matchless, Tsumeb and Kombat mines, and many others. **Zimbabwe** Common in copper deposits in the country. **Swaziland** Bornite is found at Makwamakop and Kubuta. (Chalcocite, chalcopyrite, and copper have similar distributions to bornite.)

Bornite, 15 cm. Spektakel mine, South Africa.

Brucite

Mg(OH)₂

Composition	hydroxide
Crystal system	trigonal
Hardness	2.5
Specific gravity	2.39
Streak	white
Lustre	vitreous

■ Description A soft micaceous
mineral. Brucite crystals are generally
thin, flat and tabular, but may also
be scaly, granular or needle-like.
They can be sectile, i.e. flexible
and able to bend without
breaking. Brucite occurs in a
variety of colours including
white, grey, gold, pale green
to vibrant green and grey-
blue to sea blue.

■ Occurrence
South Africa NC: Brucite
is found in the Kalahari
manganese field. LIM: Occurs
at the Palabora Mine, as well as
in the Mokopane district.

Botryoidal brucite, 3.5 cm.
N'Chwaning II mine, South Africa.

Namibia The Kombat mine has reported brucite. **Zimbabwe** Fine
specimens came from Ethel asbestos mine and others have been
reported from mines on the Great Dyke, central Zimbabwe.

Brucite, 7 cm. Palabora mine, South Africa.

Calcite
CaCO$_3$

Composition	carbonate
Crystal system	hexagonal
Hardness	3
Specific gravity	2.71
Streak	white, grey
Lustre	vitreous, pearly

Description Calcite is common, occurring in a variety of crystal forms, particularly scalenohedrons and rhombohedrons. May also occur in massive lumps, granular masses and fibrous habits, and forms stalactites and stalagmites. Colours vary and include black, white, red, yellow, green, orange, blue and brown, but it may also be colourless, transparent or opaque. Very common in limestones, dolomites, and many ore deposits.

Occurrence South Africa NC: Common from the Kalahari manganese field. LIM: Large crystals came from the Messina copper mines in the north. GAU: Occurs in veins in several Witwatersrand gold mines. KZN: Colourful calcite comes from Marble Delta. The mineral is also found in amygdales in Drakensberg basalt lavas. WC: Calvinia produced large masses.

Namibia Found in many localities, particularly the Tsumeb mine (red, yellow, green, orange, pink, white and transparent colourless crystals).

Zimbabwe Was mined west of Hwange. Also occurs in the Dorowa and Shawa carbonatites.

Swaziland A deposit is known in southern Swaziland southeast of Hluti.

Twinned calcite crystal, 6 cm. Calvinia, South Africa.

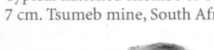

Typical flattened rhombs of calcite, 7 cm. Tsumeb mine, South Africa.

'Dog's-tooth' calcite, 4.1 cm. N'Chwaning II mine, South Africa.

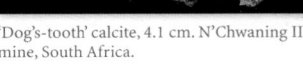

Complex calcite crystals, 8 cm. N'Chwaning II mine, South Africa.

◼ Cassiterite

SnO$_2$

Composition	oxide
Crystal system	tetragonal
Hardness	6–7
Specific gravity	6.99
Streak	white, grey-brown
Lustre	adamantine, vitreous

◼ **Description** Usually brown to brown-black, grey, yellow or black and forms lustrous prismatic crystals. It may also form botryoidal and reniform masses. Cassiterite is durable and heavy.

◼ **Uses** Cassiterite is the main economic mineral mined for tin. It is used in beneficiating the properties of other materials, e.g. solder, bronze and brass alloys. Also used in manufacturing tinfoil and some pesticides.

Cassiterite, 6.4 cm. Rooiberg mine, South Africa.

◼ **Occurrence** Common in many tin-bearing granites and particularly in pegmatites. **South Africa** NC: Deposits occur in pegmatites. LIM: Cassiterite exploited from the Bushveld Complex tin mines. KZN: Deposits known in the Umfuli area in Zululand, 15 km east of Melmoth. WC: Beautiful crystals present in the Cape granites. **Namibia** There are four 'tin belts' in central Namibia, running from a) Brandberg West to Goantagab, b) Uis to Cape Cross, c) Kohero to Nainais, and d) Omaruru to Erongo to Sandamap.

Cassiterite, 3.1 cm. Uis mine, Namibia.

Botswana Some minor deposits found in the east. **Zimbabwe** Many pegmatites exploited for cassiterite in the Hurungwe, Hwange and Mudzi districts, and others. **Swaziland** Alluvial deposits occur in Mbabane River gravels. Cassiterite is also found at Makwanakop, near the border with Mpumalanga. **Mozambique** Tin was mined at Vila Machado, northwest of Beira.

Celestine

SrSO$_4$

Composition	sulphate
Crystal system	orthorhombic
Hardness	3–3.5
Specific gravity	3.97
Streak	white
Lustre	vitreous

Description Forms characteristically heavy prismatic pale blue crystals. Cleaves easily.

Uses Celestine can be a source of strontium, which is used in the manufacture of fireworks.

Occurrence South Africa NC: Celestine has been found in the Kalahari manganese fields. Rare large blue crystals found in some kimberlite diamond mines. LIM: It is recorded at the Phalaborwa Complex and the Glenover Phosphate mine.

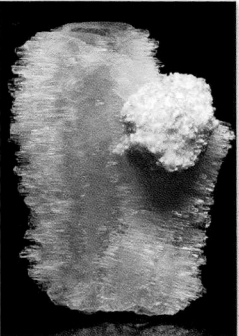

Celestine (blue) with oyelite (white), 3.8 cm. N'Chwaning II mine, South Africa.

Celestine, 5 cm. N'Chwaning II mine, South Africa.

Celestine on calcite, 5 cm. N'Chwaning II mine, South Africa.

Cerussite

PbCO$_3$

Composition	carbonate
Crystal system	orthorhombic
Hardness	3–3.5
Specific gravity	6.55
Streak	white
Lustre	adamantine, resinous

Description This heavy mineral forms beautiful snowflake-shaped, twinned crystals. These crystals may be colourless, transparent or translucent grey-white.

Reticulated cerussite, 4.1 cm. Tsumeb mine, Namibia.

Occurrence Exclusively associated with lead deposits. **South Africa** NC/MP/NW: Found in small lead deposits. **Namibia** Was a common mineral at the Tsumeb mine (which has produced the world's finest crystals). Found with other lead deposits in the Otavi mountainland and in central and southern Namibia. **Zimbabwe** Small crystals occur in the Bulawayo district, and other lead mines in the Kwekwe, Mutare and Gokwe districts. **Swaziland** Cerussite is found in the Kubuta district.

'V-twinned' cerussite coated with yellow goethite, 2.3 cm. Rosh Pinah mine, Namibia.

Reticulated cerussite, 4.1 cm. Tsumeb mine, Namibia.

Water-clear cerussite, 8 cm. Kombat mine, Namibia.

Chalcocite

Cu_2S

Composition	sulphide
Crystal system	monoclinic
Hardness	2.5–3
Specific gravity	5.5–5.8
Streak	grey, black
Lustre	metallic

See bornite, chalcopyrite and copper.

■ **Description** Common as formless lumps in hydrothermal sulphide vein deposits. Also forms steel-grey crystals, but these are rare. Chalcocite is usually associated with other copper-bearing minerals such as malachite, azurite, bornite, covellite and chalcopyrite.

■ **Uses** Chalcocite is an important copper-ore mineral.

■ **Occurrence** **South Africa** Occurs in the Bushveld Complex. NC: Fine specimens come from the Okiep copper district. LIM: Chalcocite crystals found at the Messina copper mines and, rarely, in the Phalaborwa Complex. MP: Solid masses occurred at the Stavoren tin mines. **Namibia** The Tsumeb mine produced hair-like crystals. Chalcocite is also found in the Otavi mountainland copper deposits and euhedral crystals have come from the Khan mine. **Botswana** Occurs in the northwest of the country. **Zimbabwe** The Copper Queen, Alaska and Mangula mines as well as some others have chalcocite deposits.

Chalcocite, 7.4 cm. Tsumeb mine, Namibia.

Chalcopyrite

CuFeS$_2$

Composition	sulphide
Crystal system	tetragonal
Hardness	3.5–4
Specific gravity	4.4
Streak	greenish-black
Lustre	metallic

See associated species – bornite, chalcocite and copper.

Description

Chalcopyrite is an attractive brass-yellow colour and is sometimes iridescent, resembling bornite. Commonly forms tetrahedral crystals, but may also be botryoidal, reniform or massive.

Uses

It is an important source of copper.

Occurrence

South Africa NC: Found in many copper deposits in South Africa, e.g. Messina copper mines. Aesthetic crystals are known from the Okiep copper district. It was a major ore mineral at Prieska. **Namibia** Very similar distribution to that of bornite and chalcocite – there are over 300 documented copper deposits in Namibia. **Botswana** Found at the Selebi-Phikwe nickel deposit. Also occurs in northwest Botswana and at other scattered deposits.

Chalcopyrite with apophyllite-(KOH), 4.5 cm. Belfast Granite mine, South Africa.

Chalcopyrite crystal, 1.6 cm. Nababeep West mine, South Africa.

Zimbabwe Found in most of the copper deposits with chalcocite, bornite, malachite and azurite. **Swaziland** Chalcopyrite associated with bornite at the Forbes Reef mine.

Chiastolite – see andalusite.

Copper

Cu

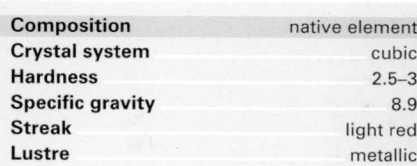

Composition	native element
Crystal system	cubic
Hardness	2.5–3
Specific gravity	8.9
Streak	light red
Lustre	metallic

See other copper species – azurite, bornite, chalcocite, chalcopyrite, chrysocolla, cuprite and malachite.

Description Occurs in several forms including flat sheets, wire-like fibres and cubic, octahedral or dodecahedral crystals. Dendritic or arborescent forms are common. Colour varies from copper-red, in fresh unoxidized material, to brown or pale pink.

Uses Copper is an excellent conductor of electricity and is used in the manufacture of copper wiring, cables, armature wiring and coils.

Occurrence South Africa NC: Large masses or finely crystalline aggregates occured in the Okiep copper mines and at Prieska. LIM: Was common in the Messina copper mines. KZN: Occurs at a few scattered localities. **Namibia** Beautiful specimens came from Tsumeb, Klein Aub, Kombat and Onganja mines. **Zimbabwe** Known from several copper mines, e.g. the Alaska, Umkondo, Mangula, Mkondo, Bikita and Falcon mines.

Copper, 7.4 cm. Otjihase mine, Namibia.

Copper with cuprite, 8.1 cm. Mufulira, Zambia.

Cordierite

$Mg_2Al_4Si_5O_{18}$

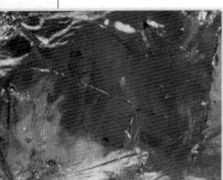

Composition	silicate
Crystal system	orthorhombic
Hardness	7–7.5
Specific gravity	2.5–2.8
Streak	white
Lustre	vitreous

Description Forms stubby and prismatic crystals, but the mineral may also be granular and massive. The crystals can be blue (most common), green, violet, grey, brown or yellow. The transparent gem variety is known as iolite or 'water sapphire'.

Occurrence Cordierite is found in granites and various metamorphic rocks. **South Africa** NC: Occurs in the gneisses of the Namaqualand Metamorphic Complex and other metamorphic terrains. **Namibia** Crystals have been found in schist and gneiss in the Namib-Naukluft Park and the Brandberg West mine. Garnet-cordierite gneiss occurs between Henties Bay and Swakopmund. **Zimbabwe** Iolite found in the Makuti area of the Hurungwe district and in the Beitbridge district. Iolite also comes from the Mutoko district.

Gem-quality cordierite, 2 cm. Zimbabwe.

Corundum

Al_2O_3

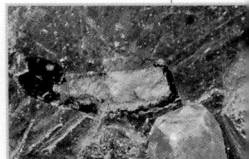

Composition	oxide
Crystal system	hexagonal
Hardness	9
Specific gravity	4.0–4.1
Streak	white
Lustre	vitreous

Description The crystals are hexagonal and barrel-shaped. Corundum is the second hardest mineral after diamond. Olive-green, brown, yellow, grey and colourless crystals occur. Transparent red crystals are the variety ruby, while transparent blue crystals are sapphire.

Uses Corundum was once used as an abrasive, but synthetic abrasives have almost entirely replaced it. Some corundum is used in the production of high-alumina refractory bricks.

Occurrence Mainly found in metamorphic rocks, but sometimes also in igneous rocks, notably pegmatites. **South Africa** NC: Ruby-red corundum occurs south of Aggeneys and in the Northern Cape province pegmatite belt. LIM: Was mined in the metamorphic rocks outcropping from Polokwane to Musina to Leydsdorp. The largest corundum crystal in the world came from this province (a grey 59 cm-long crystal, weighing 151 kg). MP: Some corundum is present in the Lydenburg and Barberton districts. KZN: Occurs near Kranskop. **Namibia** Not widespread, but is found in the Karasburg, Omaruru and Windhoek districts. **Botswana** Northeast of the Lotsane river in the Tuli Block, sapphire-blue corundum crystals (up to 3 cm) occur in alluvial deposits. **Zimbabwe** Historically, the second largest corundum producer after Russia. Blue corundum (i.e. sapphire) comes mainly from the O'Briens Prospect, Mazowe district, but also from the Beitbridge district. Occurs in several other districts, e.g. Chiredzi, Shurugwi and Rushinga. **Swaziland** Corundum deposits were exploited in the Hlatikhulu-Goedgegun region.

A selection of corundum crystals from South Africa and Zimbabwe. The central blue crystal is 4,5 cm.

Crocidolite – see riebeckite.

◼ Cuprite

$Cu_2^{1+}O$

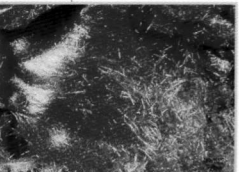

Composition	oxide
Crystal system	cubic
Hardness	3.5–4
Specific gravity	6.14
Streak	shiny brown-red
Lustre	adamantine to earthy

◼ **Description** Typically forms octahedral, cubic or dodecahedral crystals that are distinctively blood red to purple-red. If exposed to sunlight, the colour changes to metallic black. Cuprite can occur as acicular crystals called chalcotrichite.

◼ **Uses** It is an important source of copper ore.

◼ **Occurrence** **South Africa** LIM/MP: Rare, small, bright red crystals were found in some copper mines, e.g. the Messina and Stavoren mines. **Namibia** World-famous malachite-coated cuprite crystals (the largest in the world, up to 14 cm in diameter and 2.1 kg) came from the old Onganja copper mine. Tsumeb mine also contained excellent blood-red cuprite crystals and chalcotrichite. Small amounts of cuprite occur in other copper deposits, e.g. at Klein Aub, Kombat, Gorob, Copper Valley and the Natas mine. **Zimbabwe** Found in some of the oxidized portions of copper prospects in the Shurugwi, Chiredzi, Bikita, Hwange, Nkai Nkayi, Mutare and Masvingo districts.

Cuprite, mimetite and dolomite, 4.2 cm. Tsumeb mine, Namibia.

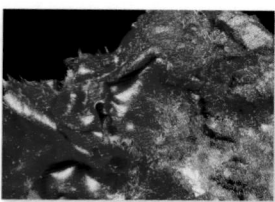

Cuprite, variety chalcotrichite, 2.2 cm. Tsumeb mine, Namibia.

Cuprite crystals, 2.3 cm. Tsumeb mine, Namibia.

Descloizite

$PbZn(VO_4)(OH)$

Composition	vanadate
Crystal system	orthorhombic
Hardness	3–3.5
Specific gravity	6.24–6.26
Streak	brown-red, orange
Lustre	greasy, lustrous

Description Forms attractive dark orange-red to brown to black crystals with a variety of shapes and habits including prismatic, stalactitic, granular or tabular. Crystals are often spear-shaped (pyramidal) and clustered together. Descloizite forms a chemical series with mottramite, another vanadium-bearing mineral.

Uses Descloizite is an important ore of vanadium, which is used in the manufacture of steel alloys, catalysts, blue and yellow pigments in paint and ceramics, electrodes in fuel cells, and as a drying agent in ink and paint. A mixture of vanadium and gallium, V_3Ga, is used as a superconductor.

Occurrence **South Africa** LIM: Rare small crystals are known near Sibasa. GAU: Light brown to dark brown crystals occur at the Argent and Edendale lead mines. **Namibia** The largest vanadium deposits in the world are found in the Otavi mountainland, notably the Berg Aukas and Abenab mines. Berg Aukas is world-famous for descloizite specimens. Descloizite also reported from Kaokoland, 20 km north of Opuwo, where it occurs, with galena, in dolomite.

Spear-shaped pyramidal descloizite, 5.2 cm. Berg Aukas mine, Namibia.

Transparent descloizite crystals, 6.8 cm. Berg Aukas mine, Namibia.

Descloizite, 4.1 cm. Berg Aukas mine, Namibia.

Diamond

C

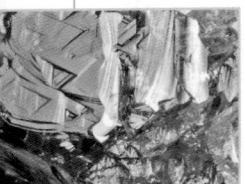

Composition	native element
Crystal system	cubic
Hardness	10
Specific gravity	3.51
Streak	white
Lustre	adamantine, greasy

Description The typical crystal form of diamond is an eight-sided octahedron, but many forms are known, including cubic crystals, dodecahedrons, spherical stones, flattened 'mackels' and combination forms. Diamond is usually colourless but variations include yellow, pink, brown, red, blue, grey, green and black. 'Bort' is a varietal name for opaque grey-black diamonds. Although it is the hardest mineral known on Mohs' Scale of hardness (10) and can scratch any other mineral, diamond can be smashed with a hammer.

Uses The most famous of all gemstones and consumed in quantity by the jewellery industry, diamonds also find applications in various tools including drilling bits, diamond-edged saw blades and super-sharp surgical scalpels.

Occurrence Diamonds are mined from their host rock, kimberlite, from alluvial river gravels, and from marine deposits off the Atlantic west coast. **South Africa** Noteworthy South African diamonds are the Cullinan (the largest diamond ever found, weighing 3 106 carats or 621,2 grams), the 997.5-carat Excelsior, the 726-carat Jonker, the 657-carat Jubilee, the 287.4-carat Tiffany, the 439.8-carat De Beers, the

Diamond crystal, 5 mm, in a matrix of the rock eclogite. South Africa.

599-carat De Beers Centenary, the 353.9-carat Premier Rose, the 426-carat Niarchos, and an unnamed 507-carat stone discovered in September 2009. **NC:** Diamonds were discovered in 1866 on the banks of the Orange River near Hopetown. They are mined from kimberlite and alluvial gravels deposited by the Vaal and Orange rivers. Noteworthy occurrences are in Kimberley at the 'Big Hole', and the Bultfontein, Dutoitspan, De Beers, Wesselton and Kimberley mines. They are also found at the Finsch mine located northwest of Kimberley. **LIM:** The Venetia mine has impressive reserves. **GAU:** The Premier mine has produced some of the largest diamonds ever found. **Namibia** Diamonds were first discovered in Namibia in 1898. All stones are recovered from marine or sedimentary deposits and no kimberlite pipes are mined for diamonds. Economic deposits occur along the southern-central Namibian Atlantic seaboard and the Orange River. **Botswana** Botswana is the world's largest diamond-producing country, mining exclusively from kimberlite pipes at Lethlhakane, Orapa and Jwaneng. **Zimbabwe**

Diamond crystal, 1 cm. South Africa.

A 138-carat diamond surrounded by smaller alluvial diamonds.

Diamond, 13 mm. A.F. Williams collection.

Kimberlites are known in Zimbabwe, but only a few contain diamonds, e.g. in the Bubi, Beitbridge (at River Ranch), Binga, Gweru and Mwenezi districts and near Zvishavane. **Swaziland** A kimberlite in the Dvokolwako area, approximately 50 km northeast of Manzini, produced diamonds. **Lesotho** Lesotho has 34 kimberlite pipes and about 140 kimberlite dykes, of which 24 contain diamonds; the Lets'eng-la-Terae kimberlite pipe contains a high concentration of large diamonds of over 10 carats, e.g. the 602-carat Lesotho Promise, the 493-carat Lesteng Legacy and an unnamed 478-carat stone found in September 2008.

Diopside

CaMgSi$_2$O$_6$

Composition	silicate
Crystal system	monoclinic
Hardness	5.5–6.5
Specific gravity	3.22–3.38
Streak	white, grey
Lustre	vitreous, dull

Description A member of the pyroxene group, diopside is a common mineral found in igneous and metamorphic rocks. The crystals are prismatic, columnar or massive and may be colourless, white or grey. Brown, green or black crystals also occur. A bright green variety, chrome-diopside, occurs in kimberlites.

Occurrence Most southern African carbonatites contain diopside and it is relatively widespread in the mafic rocks of the Bushveld Complex. **South Africa** It is a common rock-forming mineral in marbles and other metamorphosed carbonate-rich rocks. **NC:** Diopside has been reported in this province. **LIM:** Pale lime-green crystals are present at the Palabora mine (some crystals are over a metre long). **KZN:** Large crystalline masses occur at Marble Delta. **FS:** Chrome diopside is found in kimberlites, e.g. at the Jagersfontein, Roberts Victor, Monastery, and De Beers mines. **Namibia** Chrome diopside is a widespread species in a cluster of 60 kimberlites in the Gibeon-Brukkaros area, while diopside is common in skarn deposits at Ais dome, and in the Otjiwarongo district. **Zimbabwe** Chrome diopside is sometimes found as transparent crystals in kimberlites.

Unusually large, well-formed diopside crystal, 9.8 cm. Madagascar.

Dioptase
$Cu^{2+}SiO_2(OH)_2$

Composition	silicate
Crystal system	hexagonal
Hardness	5
Specific gravity	3.28–3.35
Streak	pale blue-green
Lustre	vitreous

Description Forms vibrant bright green to blue-green crystals that can be confused with emerald because of their colour, although dioptase is softer. Dioptase crystals are well formed, prismatic, can reach up to 5 cm in size in extraordinary cases, and are highly prized in mineral collections. Crystals are rare worldwide, but some of the best come from Namibia. It is a secondary copper mineral that forms at the expense of primary copper ores such as bornite or chalcopyrite.

Uses Small specimens are wrapped in silver or gold wire and used as pendants.

Occurrence **South Africa** NC: Small crystals of a few millimetres were found in the weathered gossans at Aggeneys. NW: Very rare – found at a locality near Christiana. **Namibia** The finest dioptase crystals in the world come from the Tsumeb mine and the Kaokoveld (the largest crystals measuring 5 cm, and crystals over 1 cm are common). Other localities include the Otavi mountainland and the Guchab and old Rodgerberg mines. Commonly associated minerals are azurite, malachite and calcite. **Zimbabwe** Reported from the Montana copper prospect, Chiredzi district (with malachite, chrysocolla and azurite). Also found at the Alaska copper mine, the Copper Queen zinc-copper-lead mine (west of Chinhoyi), and the Nevada, Old Mint and Cedric mines in the Makonde districts.

Dioptase and mottramite, 3.8 cm. Tsumeb mine, Namibia.

Dioptase on plancheite, 2.9 cm. Kaokoveld, Namibia.

Dioptase and malachite coated by quartz, on chrysocolla, 4 cm. Kaokoveld, Namibia.

Dolomite
CaMg(CO$_3$)$_2$

Composition	carbonate
Crystal system	trigonal
Hardness	3.5–4
Specific gravity	2.85
Streak	white
Lustre	pearly, vitreous

Description Dolomite exists as both a mineral and a rock. Crystals are rhombohedral, with distinctive, curved, saddle-shaped faces. This feature can be used to distinguish dolomite from calcite, which it closely resembles. It is white, colourless or brown, but can be coloured by the presence of trace elements such as cobalt (pink) or copper (green). Dolomite forms a chemical series with two related carbonate minerals, ankerite (iron-magnesium carbonate) and kutnahorite (calcium-manganese carbonate). Dolomite rock can dissolve in acidic water, forming underground cavities and caves such as those at Sterkfontein in Gauteng and the Cango Caves near Oudtshoorn.

Uses Dolomite is quarried for lime, which is used by the cement industry.

Dolomite, 6.8 cm. Tsumeb mine, Namibia.

Dolomite, 8 cm. Tsumeb mine, Namibia.

Occurrence **South Africa** GAU: Found as secondary vein fillings in the Witwatersrand gold mines (usually associated with quartz and pyrite), e.g. at the Kloof gold mine. NW: Saddle-shaped grey-white crystals up to 2 cm are found at the Pering mine. **Namibia** Crystals were common at the Tsumeb mine, usually grey-white, but some brown, pink, green or yellow. Rarely found from the Erongo mountains. **Zimbabwe** Common in the carbonate rocks of the Lomagundi Group, in the Darwin and Mutoko areas.

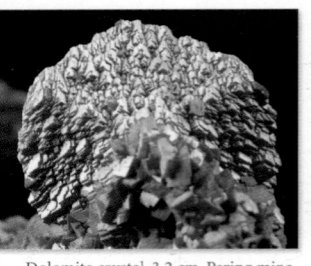

Dolomite crystal, 3.2 cm. Pering mine, South Africa.

Elbaite

$Na(Li,Al)_3Al_6(BO_3)_3Si_6O_{18}(OH)_4$

Composition	silicate
Crystal system	hexagonal
Hardness	7
Specific gravity	3.03–3.10
Streak	white
Lustre	vitreous

Description Elbaite is a member of the tourmaline group of minerals. It has a complex chemical composition and is multicoloured. The different colours are caused by the presence of trace elements. Elbaite crystals are prismatic and elongate, with typical striations or grooves on the outer crystal faces that run parallel to the long axis of the crystal. The mineral is common in granites, particularly granitic pegmatites. The blue variety of elbaite is called indicolite and the red, rubellite. 'Watermelon tourmaline' has a green outer rim and a pink-to-red-zoned core.

Occurrence **South Africa** Gem-quality elbaite is rare in South Africa. NC: The pegmatites at Straussheim and Angelierspan contain bright blue, green and pink elbaite crystals. At Norrabees, north of Steinkopf, some concentrically zoned green and pink elbaites have been found. **Namibia** Tourmaline is one of the most important gemstones found in Namibia. It is concentrated in tin-lithium-beryllium pegmatites from the Karibib-Usakos area (where excellent rubellite and indicolite are both exploited). It also occurs in the Sandamap-Erongo area, and in pegmatites near Uis and the Brandberg.

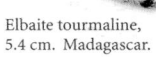
Elbaite tourmaline, 5.4 cm. Madagascar.

Zimbabwe Rubellite tourmaline has been produced from the Mwami pegmatites. Green tourmaline is more common from the Mwami region and the pegmatites in the northeastern region. St Ann's pegmatite has produced half of Zimbabwe's gem tourmaline. **Swaziland** A pegmatite near Kubuta contains rubellite tourmaline with lepidolite.

Elbaite tourmaline, 2.8 cm.
Karibib district, Namibia.

Elbaite tourmaline, 4 cm. Karibib district, Namibia.

Epidote

$Ca_2(Fe^{3+},Al)_3(SiO_4)_3(OH)$

Composition	silicate
Crystal system	monoclinic
Hardness	6–7
Specific gravity	3.35–3.5
Streak	white, grey
Lustre	vitreous

Description Epidote is found as light green to dark black-green prismatic crystals. It varies from transparent to opaque black (in larger crystals). The most common crystal habits are smooth or striated, elongate and prismatic, but epidote can also be needle-like, fibrous or massive. Chemical impurities can cause it to take on different colours, e.g. manganese produces piemontite (a red-orange colour). A famous variety – vibrant mauve-blue tanzanite – is mined in Tanzania. The variety unakite is a mixture of green epidote and pink orthoclase feldspar.

Uses The variety tanzanite is a valuable gemstone.

Occurrence **South Africa** NC: A large epidote deposit occurs near Neilersdrif between Kakamas and Keimoes. Unakite is found in this area. LIM: Good quality unakite occurs here, while both green epidote and piemontite are found at the Messina copper mines. **Namibia** At Nauchas, in the Gamsberg region of the Rehoboth district, beautiful lustrous dark green to black crystals occur. **Botswana** Epidote is fairly widespread in mafic schists in the Tati schist belt. **Zimbabwe** The mineral occurs in the Rushinga, Masvingo and Hurungwe districts and lapidary grade unakite is exploited in the Beitbridge district. **Swaziland** It occurs in greenstone belt rocks in the northwest. At Makwanakop, an epidote-rich rock contains malachite, bornite and chalcopyrite.

Epidote, 4.9 cm. Northern Cape, South Africa.

Epidote, variety piemontite, 2.1 cm. Messina mine, South Africa.

Ettringite

$Ca_6Al_2(SO_4)_3(OH)_{12} \cdot 26H_2O$

Composition	sulphate
Crystal system	hexagonal
Hardness	2–2.5
Specific gravity	1.77
Streak	white
Lustre	vitreous

Description Generally forms beautiful hexagonal, prismatic bright yellow crystals, but other shades include amber, brown, orange-yellow, canary yellow, yellow-green and white. It tends to be somewhat unstable and to dehydrate, alter and change colour with time. Can be confused with sturmanite.

Uses The vibrant yellow colour makes ettringite very popular with collectors.

Occurrence South Africa

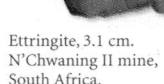

Ettringite, 3.1 cm. N'Chwaning II mine, South Africa.

NC: Ettringite comes exclusively from the Kalahari manganese field mines, most notably from N'Chwaning II mine (individual crystals over 15 cm are known). Many crystals were found in the 1980s but it has been relatively scarce since then. South African ettringite crystals are the finest in the world because of their large size and beautiful colour. The species is rare worldwide, but the initial discovery of ettringite in the Kalahari manganese fields in the 1980s produced thousands of specimens. It has never again been found in such quantities, but very beautiful specimens have been found sporadically during recent years.

Ettringite and calcite matrix, field of view 3.8 cm. N'Chwaning II mine, South Africa.

Ettringite on calcite, 4.8 cm. N'Chwaning II mine, South Africa.

Feldspar group – see microcline and orthoclase.

Ferberite

Fe^{2+}WO$_4$

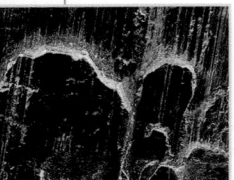

Composition	tungstate
Crystal system	monoclinic
Hardness	4–4.5
Specific gravity	7.51
Streak	black to brown-black
Lustre	submetallic to adamantine

Description Ferberite (previously known as 'wolframite') forms heavy prismatic crystals. It is found in granitic pegmatites and may be black, brown, red-brown or metallic grey.

Uses Ferberite contains tungsten, which has the highest melting point of all metals: 3 410 °C. It is an extremely hard yet tensile metal. Tungsten carbide is used in drill bits and abrasives and in tool-and-die apparatus. Electric light bulb filaments are made of tungsten.

Occurrence **South Africa** NC: Many small tin-tungsten deposits occur in the province. Found in the area around Springbok, Okiep and Nababeep (in quartz veins in schists that contain minor amounts of molybdenite, chalcopyrite, and bismuth-bearing minerals). Tungsten deposits occur in a broad belt along the Orange River between Vioolsdrif and Upington. MP: Ferberite comes from the Bushveld Complex, notably the Mutue Fides and Stavoren deposits. **Namibia** Ferberite is found in pegmatites and quartz veins. Substantial amounts occurred at the Krantzberg tungsten mine, northeast of the Erongo mountains (associated with titanite, tourmaline, topaz, quartz and fluorite). **Zimbabwe** The mineral is found mainly in granites and in metamorphic rocks in the northern and northeastern parts of the country. The Honey Mine had well-formed large ferberite crystals. **Swaziland** Ferberite is reputed to come from Mtshengu's drift on the Mkondo River near Goedgegun.

Ferberite, 3.2 cm. Northern Cape, South Africa.

Ferberite, 5 cm. Northern Cape, South Africa.

Ferrohornblende

$Ca_2Fe_3^{2+}[(Al,Fe^{3+})]Si_7AlO_{22}(OH)_2$

Composition	silicate
Crystal system	monoclinic
Hardness	5–6
Specific gravity	3.41
Streak	grey-white
Lustre	vitreous

■ Description A member of the amphibole group, this mineral forms prismatic elongate black crystals. The crystal faces are striated.

■ Occurrence Some granites may contain ferrohornblende. A relatively common mineral in mafic igneous and metamorphic rocks such as basalt, gabbro, syenite, amphibolite, alkaline rocks, and calc-silicates distributed throughout **South Africa** (particularly in the Bushveld Complex, Limpopo Belt and Namaqualand), **Namibia** (in the Matchless amphibolite belt), **Botswana** and **Swaziland**.

Fine layers of black ferrohornblende in gneiss, 15 cm, Margate, South Africa.

Black ferrohornblende crystals with quartz in amphibolite, 5.5 cm. South Africa.

Fluorite

CaF$_2$

Composition	halide
Crystal system	cubic
Hardness	4
Specific gravity	3.18
Streak	white
Lustre	vitreous

Description Crystals are characteristically simple cubes, but octahedral, dodecahedral and combination forms exist. Occurs in a confusing array of colours, displaying almost all shades, including white, yellow, orange, green, brown, red, pink, blue, purple, black and colourless. It often fluoresces under ultraviolet light.

Uses A major economic mineral mined for fluorine, which is used in the metallurgical and chemical industries (particularly in the production of hydrofluoric acid). It is also used as a flux and is added, somewhat controversially, to drinking water.

Occurrence Fluorite occurs in acidic igneous rocks and certain sedimentary rocks, notably limestone and dolomite. **South Africa** NC: Fluorite was found in the Okiep district. LIM: Unusual crystals occur in the Pilanesberg. It is plentiful at the old Bushveld Complex tin mines. Found as veins and irregular bodies in the host rocks of the Rooiberg-Bela-Bela area. GAU: The largest deposit of fluorite in the world occurs at the Vergenoeg fluorspar mine northeast of Pretoria. NW: Relatively common from deposits in the province. KZN: Localities include the

Fluorite, 7.2 cm. Okorusu mine, Namibia.

Aladdin fluorite mine, Sinkwazi on the north coast and the area south of Nongoma. **Namibia** The Okorusu deposit 48 km north of Otjiwarongo produces beautiful cubic green, purple and yellow crystals. Vibrant green crystals several centimetres long occur in the pegmatites in the Erongo mountains, at Klein Spitzkoppe and at other scattered deposits. **Botswana** Found in quartz veins (with galena and chalcopyrite) at the Bushman mine, 60 km southeast of Nata and in the Gaberone granite at Ditshutswane, between Kika and Manyana. Fluorite is known from west of the Taupse River.

Fluorite, 7.1 cm.
Okorusu mine, Namibia.

Zimbabwe Found in the Hwange and Hurungwe districts. Pale green and dark purple fluorite are found in quartz veins in gneiss at the Marion mine. Several other small deposits occur as well.
Swaziland Two deposits are found in the Hlatikhulu district.
Mozambique Has been exploited at Serra Gorungoza (at two localities close to the northeastern border with Zimbabwe).

Fluorite on quartz, 5.2 cm. Buffalo Fluorspar mine, South Africa.

Fluorite on quartz, 13.5 cm. Riemvasmaak, South Africa.

Galena

PbS

Composition	sulphide
Crystal system	cubic
Hardness	2.5
Specific gravity	7.58
Streak	grey
Lustre	metallic

Description Galena is steel grey, with a distinctive metallic lustre, and is characteristically very heavy because of its lead content. The most common crystal form is cubic, although octahedrons and combination forms exist. Cleaves easily.

Uses It is the main economic ore mineral of lead and was used as a petrol additive, although unleaded petrol is now the norm. Used to make ammunition, brass and bronze alloys and as an additive in pesticides.

Occurrence Relatively common mineral in base metal, gold and other metallic metamorphic and hydrothermal deposits. **South Africa** Known from many localities in dolomite. NC: Occurs at Bushy Park, Griquastad district, and at the Broken Hill mine, Aggeneys. LIM: Leeuwbosch mine produced good specimens. NW: Found at Pering mine, Bokkraal 344 JP, Rhenosterhoek 343 JP and the old Doornhoek lead mine. **Namibia** Galena occurs at the Ai-Ais mine in southern Namibia, the Rosh Pinah mine and the Namib Lead mine. It was one of the main ore minerals at the Tsumeb mine. **Botswana** Occurs at the Lady Mary mine, in eastern Botswana and the Bushman mine, southeast of Nata. **Zimbabwe** Most galena is associated with gold deposits in the ancient greenstone belts, with lead as a by-product. **Swaziland** Galena (with cerussite, malachite and bornite) is found in quartz veins at Kubuta. A sulphide-mineralized quartz vein occurs in outcrops in the Ngwavuma valley, east of Hlatikhulu.

Galena (large cubic grey crystals) on dolomite, 5.2 cm. Pering mine, South Africa.

Garnet group – see almandine, andradite, grossular, pyrope and spessartine.

Goethite

α-Fe^{3+}O(OH)

Composition	hydroxide
Crystal system	orthorhombic
Hardness	5–5.5
Specific gravity	3.3–4.3
Streak	brown, orange
Lustre	dull, submetallic

See hematite.

Description Goethite is usually ochre to orange-red because it forms from the oxidation of iron-bearing minerals, e.g. hematite, magnetite, pyrite and other iron oxides and sulphides. It is most common as a soft, earthy, microcrystalline coating on rocks and other minerals. However, occasionally it forms clusters of crystals enclosed in quartz.

Occurrence South Africa The most common iron mineral in many orebodies, where it imparts a reddish coloration in hand specimens. NC: Very attractive goethite/hematite pseudomorphs after pyrite were found from the N'Chwaning II mine, Kalahari manganese field. LIM: Forms minute golden crystals included in quartz crystals at Musina. MP: Large pseudomorphs are found in the Lydenburg district near Mt Anderson. GAU: Common at the Vergenoeg fluorspar mine (as black or iridescent stalactitic masses in vugs and caves). Goethite pseudomorphs after pyrite occur at Monument Park in Pretoria. **Namibia** Found in large amounts on the farm Eisenberg 78, Otjiwarongo district. Was abundant at the Berg Aukas mine, and at Okorusu fluorite mine. Widespread in the iron-rich rocks in the Otjosondu manganese field. Pseudomorphs of goethite after siderite are found in the pegmatites at Klein Spitzkoppe, the Erongo mountains and Onganja mine. **Zimbabwe** Goethite is associated with the weathering of banded-iron formation rocks, e.g. in the Mwanesi range. **Swaziland** Beautiful golden microcrystals, sometimes aggregated into circular haloes included in quartz, come from the Devil's Reef gold deposit, Piggs Peak district. Goethite was a very common constituent of the iron ore deposits at Ngwenya.

Iridescent goethite, 9.5 cm. Vergenoeg mine, South Africa.

Gold
Au

Composition	native metal
Crystal system	cubic
Hardness	2.5–3
Specific gravity	19.3
Streak	gold
Lustre	metallic

Description Gold is a very heavy, soft and malleable metal when it occurs in its pure form. Crystals are rare (usually cubic or octahedral). More usually it is found in the form of thin sheets and plates, or alluvial nuggets. Tiny grains may also be scattered in gold-bearing rocks.

Uses Gold has had a high monetary value since antiquity and is probably the best-known precious metal on Earth. Being malleable, it can be hammered, stretched and rolled without breaking and it doesn't rust, which makes it ideal for use in jewellery.

Occurrence Most of southern Africa's gold occurs in three geological environments, i.e. in ancient placer deposits, in shear zones and in pegmatites. **South Africa** Gold is synonymous with southern Africa, particularly South Africa, although production has steadily decreased in the twenty-first century. NC: Deposits once occurred at Hopetown and Calvinia. LIM: Historically gold was found at Eersteling and Murchison. Beautiful gold beads, gold wire, amulets and gold animal artefacts have been exhumed from Mapungubwe, west of Musina, and at Thulemela in the Kruger National Park. MP: Pilgrim's Rest and Barberton were once worked for gold. GAU: The Witwatersrand Supergroup quartz pebble conglomerates comprise the largest gold deposit on Earth. Most gold found here is microscopic – large Witwatersrand gold specimens are rare and valuable. WC: Historically it was found in the Prince Albert district, at Bredasdorp (alluvial deposits), in the Swellendam district, near Heidelberg (in quartz veins), in the Kragga River near Worcester, at Montagu, Ceres and Somerset West, and on Lion's Head overlooking Cape Town.

Gold in quartz, 4 cm. Welkom goldfield, South Africa.

Gold in conglomerate, 5.5 cm. Witwatersrand goldfield, South Africa.

EC: Historic deposits occurred in the province, specifically around Knysna. **Namibia** First discovered in Namibia in 1899 close to Rehoboth, there are many small deposits in the country. Most of the gold mined comes from sulphide minerals that contain trace amounts of the metal. Important localities include the Navachab mine in the Karibib district, the Ondundu goldfield in Damaraland and the Omaruru district (where it is found as alluvial gold). **Botswana** It is restricted to the east, e.g. the Tati schist belt, where gold is found in quartz veins. Approximately 60 abandoned gold mines and diggings occur here, all located in greenstone belt rocks. **Zimbabwe** Thousands of gold deposits, although many are very small, occur in the ancient Archaean rocks (particularly the greenstone belts). Important mines include the Cam and Motor mine, Kadoma district, the Globe and Phoenix mine, Kwekwe district, the Rezende mine, Mutare district, the Shamva mine, Shamva district, and the Dalny mine, Chegutu district.

Swaziland Gold deposits and mines are found along the northwestern border, between Oshoek and Horo, in the rocks of the Barberton greenstone belt. Mines operated in the Forbes Reef area and at Piggs Peak. The Devil's Reef mine was particularly rich.

Mozambique Alluvial gold deposits occur along the Revue River close to Manica at Bragança, and at Monarch mine near the Zimbabwe border.

Gold in

Graphite
C

Composition	native element
Crystal system	hexagonal
Hardness	1–2
Specific gravity	2.1–2.23
Streak	black
Lustre	earthy, dull, metallic

Description Usually forms black to silvery-black platy crystals and flakes. It is characteristically soft and ductile, allowing the crystals to bend without breaking. Graphite and diamond are both composed of carbon, but the way the atoms bond together means that the former is soft and friable, the latter extremely hard.

Uses Used in pencil lead and in the manufacture of brake linings, carbon bushes, batteries, crucibles, and refractory bricks.

Occurrence South Africa Occurs in regions with metamorphosed limestones and gneisses, and where some coals have been metamorphosed. NC: Found on the farm Oup 80, in Namaqualand. LIM: Occurs at the Mutale and Gumbe mines, east of Musina. KZN: Found in graphitic schists at Kwa-Xolo and in white marble at Umzimkulu. **Namibia** Small flakes are present in a deposit 20 km southwest of Otjiwarongo. Graphite occurred 60 km east of the town of Aus as pods and lenses in granite. **Botswana** Known from dolomites and graphitic schists at Moshaneng, and near the Bushman mine/ Phudulooga area. **Zimbabwe** Graphite is found in high-grade metamorphic rocks, such as gneisses, in the Hurungwe region. The Lynx mine was the main producer. Other deposits in the district include Graphite K̲i̲ Juma and Silak

Graphite, 5.5 cm. South Africa.

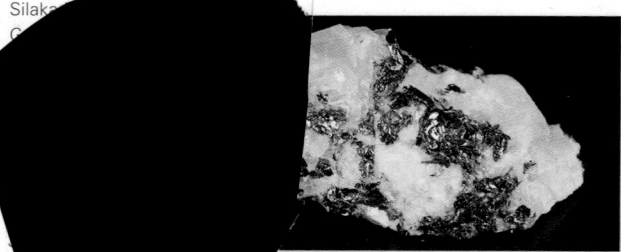

calcite, 6 cm. Marble Delta, South Africa.

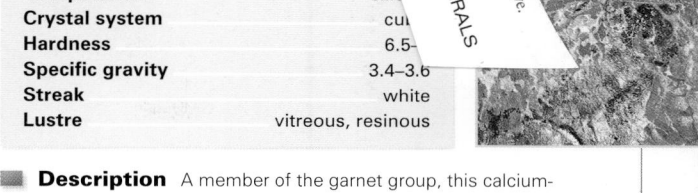

QUARTZ. 4.4 cm. Zimbabwe.

MINERALS

Composition	silic...
Crystal system	cu...
Hardness	6.5-
Specific gravity	3.4–3.6
Streak	white
Lustre	vitreous, resinous

Description A member of the garnet group, this calcium-bearing species can form beautiful orange, red, yellow or green dodecahedral crystals. Hydrogrossular garnet, known as 'Transvaal jade', is a popular green, pink, red or even translucent lapidary variety. Because it is finely crystalline it polishes well.

Uses The variety 'Transvaal jade' is used as a gemstone.

Occurrence Grossular commonly occurs in close association with chromitite seams. **South Africa** NC: Found at a few localities in the province. LIM: Known from the Steelpoort, Mokopane and Letaba districts. GAU: 'Transvaal Jade' occurs in the Bushveld Complex, Rustenburg district, approximately 70 km west of Pretoria (in close association with chromite ore) but is now mostly worked out. KZN: Gemmy orange to red grossular is found at Marble Delta. **Namibia** Grossular comes from the Ais dome skarn deposit in the Omaruru district. **Zimbabwe** Relatively widespread as brown crystals in pegmatites, calc-silicate rocks, granites and metamorphic equivalents.

Hydrogrossular garnet ('Transvaal jade'), 11 cm. Brits district, South Africa.

Grunerite

$(Fe^{2+},Mg)_7Si_8O_{22}(OH)_2$

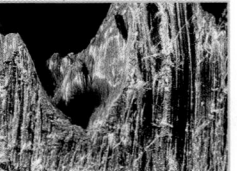

Composition	silicate
Crystal system	monoclinic
Hardness	3.5–4
Specific gravity	3.77
Streak	blue
Lustre	vitreous

Description A member of the amphibole group, this mineral forms soft white-silver asbestiform fibres over 20 cm long, and occurs in layers or seams. It is also known as amosite asbestos.

Occurrence **South Africa** The asbestos seams are layered and continuous, extending over long distances. **LIM:** Found in banded-iron formations at Penge. **NW:** Grunerite is found southeast of Zeerust.

Grunerite asbestos, 31 cm. Penge mine, South Africa.

Very long fibres of grunerite asbestos. Penge mine, South Africa.

Gypsum
$CaSO_4 \cdot 2H_2O$

Composition	sulphate
Crystal system	monoclinic
Hardness	2
Specific gravity	2.32
Streak	white
Lustre	vitreous, pearly

Description Relatively common in arid regions in southern Africa, where evaporation causes crystallization of gypsum in the soil. A soft colourless to white mineral, it cleaves easily. 'Desert rose' is a variety of gypsum resembling rose petals: the interlocking crystal blades contain sand and clay. Selenite is a semi-transparent to transparent euhedral gypsum.

Uses Gypsum is used in the manufacture of builder's plaster, gypsum boards and related construction materials.

Occurrence **South Africa** A common mineral in many saltpans and some ore deposits. NC: Occurs northeast of Kimberley and at Winsorton. Sword-like crystals are known from the Okiep copper mines and large transparent colourless crystals are present at the Kalahari manganese field. KZN: Desert roses come from the Tugela Valley, Greytown district. WC: West of Vanrhynsdorp desert roses and selenite occur in pans. **Namibia** The coast from Lüderitz to north of Walvis Bay is famous for desert roses. In addition, large, well-formed crystals have come from the Tsumeb mine and Rosh Pinah. **Botswana** The Foley gypsum deposits are found 50 km west of Tonota and in the surrounding areas. Gypsum layers also occur 40 km north of Ngware and at Sua Pan. **Zimbabwe** Gypsum is present on Nottingham Farm in the Beitbridge district, in the Gokwe and Sebungwe districts, and at the Mangula copper and Antelope asbestos mines.

Gypsum crystal, 14.2 cm. N'Chwaning II mine, South Africa.

Gypsum, variety desert rose, 8.2 cm. Lüderitz, Namibia.

Gypsum, variety desert rose, 5.8 cm. South Africa.

Halite
NaCl

Composition	halide
Crystal system	cubic
Hardness	2
Specific gravity	2.17
Streak	white
Lustre	vitreous

Description Halite – common rock salt – is colourless to white and forms soft cubic crystals that are soluble in water. Forms by evaporation in saline pans in arid areas, particularly on the coast of Namibia. Algae in the water may colour halite pink or green.

Uses Most commonly used as a seasoning on food. It is also employed in the manufacture of chlorine, caustic soda, paper, glass, soap, detergents and for food preservation. Along the coast north of Swakopmund, Namibia, it is used to surface some roads.

Occurrence **South Africa** NC/NW: Found in saltpans in low-rainfall regions from Calvinia to Kimberley and north to Vryburg. Some large saltpans occur about 100 km north/northwest of Upington. WC/EC: Coastal saltpans north and east of Cape Town contain halite. Saltpans occur around Mossel Bay and Port Elizabeth, e.g. at Koega. **Namibia** Has a somewhat similar distribution to gypsum, i.e. along the coastline between Walvis Bay and Cape Cross. Saltpans similar to those in the arid regions of South Africa occur in Ovamboland, north of Etosha Pan. Commercial salt-mining operations close to Swakopmund artificially maintain over 30 evaporation pans where halite crystallizes and is harvested. **Botswana** Found in evaporation pans in arid regions. Has been commercially mined at Sua Pan, east of Maun. **Zimbabwe** Scarce – occurs at mineral springs in the Wankie and Sebungwe districts and in the Sabie Valley.

Halite, 5.5 cm. Swakopmund, Namibia.

Hausmannite

$Mn^{2+}Mn_2^{3+}O_4$

Composition	oxide
Crystal system	tetragonal
Hardness	5.5
Specific gravity	4.84
Streak	brown
Lustre	metallic

Description Forms tetragonal black to silver-black crystals with a distinctive brown streak. Crystals possess a variety of different habits. The most distinctive is the pyramidal pagoda-like form of many hausmannite specimens from the manganese mines in South Africa.

Uses Manganese metal is used extensively in the manufacture of ferromanganese and other alloys, in dry-cell batteries, as well as in the chemical industry.

Occurrence Hausmannite is rare, but some of the finest crystals known come from the region. **South Africa** Most commonly found in the Kalahari manganese field. NC: Hausmannite crystals from Wessels and N'Chwaning I and II mines are arguably the world's best (up to 10 cm on edge, jet black and highly lustrous). **Namibia** Found at the Kombat mine, Otavi mountainland. The largest deposit of hausmannite in Namibia occurs in the Otjosondu manganese field, 150 km northeast of Okahandja.

Hausmannite, 7.4 cm.
N'Chwaning II mine, South Africa.

Hausmannite on andradite garnet, 3.7 cm. N'Chwaning II mine, South Africa.

Hematite

α-Fe₂O₃

Composition	oxide
Crystal system	hexagonal
Hardness	5–6
Specific gravity	5.26
Streak	red to brownish red
Lustre	metallic, submetallic

Description Hematite, an iron oxide mineral, is relatively common and is dispersed in a variety of rocks. It can occur in a platy, micaceous habit referred to as specularite or specular hematite, or as banded-iron formations (i.e. where entire rocks comprise very finely disseminated hematite grains). Beautiful bright silver crystals also sometimes occur. Hematite colours rocks red.

Uses Iron is used to manufacture steel and related ferroalloys, ferromanganese and ferrosilicon. It is also used as a traditional cosmetic in the form of red ochre.

Occurrence Most common and abundant in sedimentary banded-iron formations in southern Africa. **South Africa** NC: Specularite and ochre occur in caves in the province, with extensive workings found northwest of Postmasburg. Sishen-Beeshoek districts have huge iron ore deposits. The Kalahari manganese field produces beautiful large silver hematite crystals. GAU: Attractive reniform masses are found at Vergenoeg fluorspar mine northeast of Pretoria. NW: Blood-red hematite

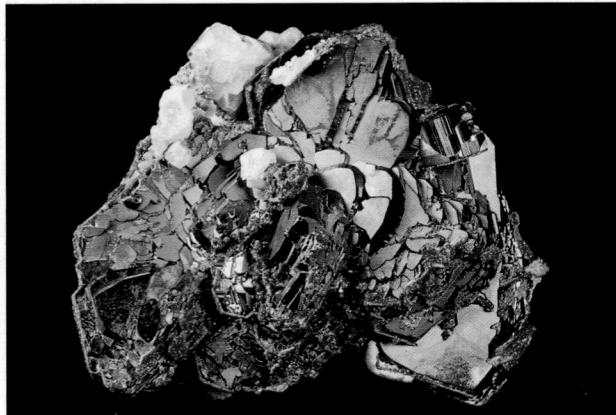

Hematite with barite and andradite garnet, 10.5 cm. Wessels mine, South Africa.

included quartz occurs along the Orange River. **Namibia** Hematite has been mined from syenites and carbonatites in the Kalkveld Complex. The Goboboseb mountains have produced water-clear quartz crystals with inclusions of tiny bright red hematite flakes. North of the Orange River hematite is found included as microscopic red crystals in quartz from pegmatites. **Botswana** Occurs in banded-iron formations outcropping in the Tati schist belt and in the southeast (south of Kanye and southeast of

Hematite pseudomorph after pyrite, 3.2 cm. Devil's Reef, Swaziland.

Lobatse). **Zimbabwe** Hematite has been mined here for aeons and used to produce a variety of iron-age tools, weapons and implements in the Kwekwe, Gweru, Chiredzi and Masvingo districts, among others. Reniform hematite came from the Yank mine, Kadoma district. **Swaziland** Ngwenya, 25 km northwest of Mbabane, was the largest iron ore deposit in the country. Attractive pseudomorphs of hematite after pyrite come from the Devil's Reef area and from Iron Hill, 2 km south of the Havelock mine in the Piggs Peak district. Hematite also exists at Gege and Maloma.

Hematite, 4.5 cm. Wessels mine, South Africa.

Inesite

$Ca_2Mn_7^{2+}Si_{10}O_{28}(OH)_2 \cdot 5H_2O$

Composition	silicate
Crystal system	triclinic
Hardness	5.5
Specific gravity	3.03
Streak	white
Lustre	vitreous, silky

■ **Description** Inesite is named for its fibrous appearance – the word *ina* is Greek for 'fibre'. This mineral commonly forms acicular, hard, fibrous crystals that occur in aggregates. It is generally pink to red, and rare transparent platy crystals also occur.

■ **Occurrence** South Africa NC: Some of the finest inesite crystals in the world came from the Wessels Mine in the Kalahari manganese field. These included both aggregates of sharp, radiating, tabular, transparent red crystals (up to 1 cm wide), and dense spherical aggregates (up to 2 cm in diameter) grouped together like grapes. During June 1996, thousands of inesite specimens were collected from a brecciated fault zone at the N'Chwaning II mine. These typical, needle-like crystals coated or partially coated the rock fragments. Most specimens were dull pink to cinnamon-brown, but some were an attractive dark pink colour.

Inesite, N'Chwaning II mine, Kalahari manganese field, South Africa.

Inesite, 8.6 cm. N'Chwaning II mine, South Africa.

Kaolinite

$Al_2Si_2O_5(OH)_4$

Composition	silicate
Crystal system	triclinic
Hardness	2–2.5
Specific gravity	2.6
Streak	white
Lustre	earthy, dull

Description This clay mineral is white and powdery, solid or granular. It forms from the chemical breakdown of other minerals such as alkali feldspars. Kaolinite originates in several geological environments and is a common constituent in the matrix of clastic sedimentary rocks, particularly sandstones.

Uses Used in manufacturing porcelain and tiles, paper, paint and plastics, in insecticides, as a carrier in fertilizer, and as a medicinal compound for treating diarrhoea. It also produces the glossy finish on paper.

Occurrence **South Africa** GAU: Deposits are known east of Benoni and in the Westonaria district, as well as about 40 km northeast of Pretoria. WC: The large deposits around the Cape Peninsula, e.g. at Noordhoek, Vredendal and Saldanha Bay, formed from weathered granites. EC: Kaolinite occurs near Grahamstown. **Botswana** An economic deposit of kaolinite is found in the Makoro area of southeastern Botswana (formed from the breakdown of alkali feldspars in sandstones).
Zimbabwe Exploited in several localities in the Mutare, Hwange, Nkayi, Mazowe and Harare districts. **Swaziland** Deposits occur in the Mahlangatsha and Mankaiana districts (formed by the breakdown and alteration of potassium feldspar).

Kaolinite, 11 cm. South Africa.

Kyanite

Al_2SiO_5

Composition	silicate
Crystal system	triclinic
Hardness	4–7.5
Specific gravity	3.53–3.67
Streak	white
Lustre	pearly, vitreous

Description Usually forms elongate, bladed, translucent to opaque crystals, with a characteristic blue colour, although rare colour variations are known. It may also occur in clusters. Kyanite, sillimanite and andalusite are common in metamorphic rocks and are polymorphs, i.e. they are chemically identical, differing only in their crystal systems.

Uses Used as a refractory mineral. It is able to withstand very high temperatures of over 1 500 °C.

Occurrence **South Africa** NC: Euhedral crystals have been found near Port Nolloth. NW: Commercial deposits are associated with shales in the Bushveld Complex aureole, e.g. the Marico district. KZN: South of Nkandla, rare gem crystals occur. FS: The Vredefort district has blue kyanite crystals. **Namibia** Translucent crystals occur in quartz veins in schist near the old Gorob mine, Namib Desert Park. There are several other kyanite localities in the Windhoek and Rehoboth districts. **Botswana** Commercially mined at Halfway Kop, 15 km southeast of Francistown. **Zimbabwe** Deposits are known from metamorphic terrains, especially in the Mwami area, northeast of Karoi. At Chipungwe, in the Hurungwe district, kyanite pseudomorphs after chiastolite occur in graphite schists. At 'Kyanite Hill', Rushinga district, attractive gem-quality blue kyanite is found. Several kyanite deposits are exploited east of Mutoko, near the Mozambique border. **Mozambique** Kyanite has been mined at Nhazonia, close to the eastern Zimbabwe border, 200 km northwest of Beira.

Kyanite crystal, 5 cm. Zimbabwe.

Kyanite crystals of various colours, field of view 8 cm. Port Nolloth, South Africa.

Lepidolite

$K(Li,Al)_3(Si,Al)_4O_{10}(F,OH)_2$

Composition	silicate
Crystal system	monoclinic
Hardness	2.5–3
Specific gravity	2.8–3.3
Streak	white
Lustre	pearly

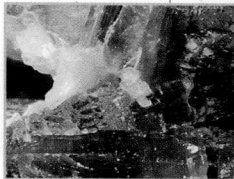

Description Lepidolite is a member of the mica group of minerals. The platy, micaceous, hexagonal lilac to mauve crystals may be several centimetres in diameter. Their attractive colour is caused by the presence of lithium.

Uses Lithium has uses in air conditioners, welding fluxes, bleaches and sanitary agents. It is added to glass and ceramics to increase the strength of these materials and to reduce their thermal expansion properties. Lithium carbonate ($LiCO_3$) is used medicinally as a treatment for depression.

Occurrence South Africa NC: Occurs in pegmatites with other lithium species, including spodumene, petalite and amblygonite. MP: Lithium-bearing pegmatites outcrop a few kilometres west of the Swaziland border. **Namibia** Found in pegmatites in the Karibib-Usakos area and the Warmbad district. Also extracted from tin-bearing pegmatites north and northwest of Karibib in the Cape Cross-Uis belt. **Botswana** The Prospect mine west of Francistown is a source of lepidolite. **Zimbabwe** A major global producer of lithium, nearly all mined from one pegmatite – Bikita, east of the Masvingo greenstone belt. Also mined from other pegmatites in the Mberengwa district. Beautiful purple crystals were collected at the Mauve mine, Harare district. **Swaziland** Fine-grained lepidolite is found in a pegmatite near Kubuta (with rubellite tourmaline crystals up to 12 cm long).

Mauve lepidolite with albite feldspar and red elbaite tourmaline, 6 cm. Muiane mine, Mozambique.

Magnetite

$Fe^{2+}Fe_2^{3+}O_4$

Composition	oxide
Crystal system	cubic
Hardness	5.5–6.5
Specific gravity	5.18
Streak	black
Lustre	metallic, dull

Description A member of the spinel group, magnetite most commonly forms octahedral black to metallic silver crystals. Its chief diagnostic feature is that it is magnetic. Magnetite is named after the Greek shepherd, Magnes, reputed to have discovered the mineral when it stuck to the iron nails in his shoes and the iron ferrule of his staff.

Uses Magnetite can be a source of iron, as well as titanium and vanadium.

Occurrence South Africa A few hills in the Bushveld Complex are composed entirely of magnetite. NC: Abundant and widespread, it is disseminated in igneous and metamorphic rocks, e.g. in banded-iron formations at Aggeneys (where well-formed octahedral crystals occur). LIM: Very common in the rocks of the Phalaborwa Complex.
Namibia Titaniferous magnetite occurs in the Kunene Complex gneisses. Goethite/hematite pseudomorphs after magnetite are found in the Erongo pegmatites. Magnetite quartzites are widespread in metamorphic terrains, e.g. in the Matchless amphibolite belt.
Zimbabwe Common in many different rock types, particularly in carbonatites such as at Shawa and Dorowa. Also common in many of the country's asbestos deposits. **Swaziland** Occurs together with hematite, goethite and siderite at Ngwenya iron mine. Found in quartzites, with goethite, at Gege and Maloma and at an iron ore deposit on the Mhlatuze River.

Flattened magnetite crystals,
2.4 cm. Palabora mine, South Africa.

Magnetite crystals, 3.4 cm. Aggeneys mine,
South Africa.

Malachite

$Cu_2^{2+}(CO_3)(OH)_2$

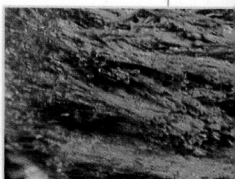

Composition	carbonate
Crystal system	monoclinic
Hardness	3.5–4
Specific gravity	4.05
Streak	pale green
Lustre	vitreous

Description Malachite is an attractive dark green mineral. Crystals are commonly acicular, but the mineral can form botryoidal masses and stalactites or stalagmites. It also occurs as green encrustations on rocks associated with copper deposits.

Uses Popular as a gemstone, lapidary material and collector's item.

Occurrence A widespread secondary oxidation mineral found in copper-rich ore deposits, often associated with azurite and common as malachite pseudomorphs after azurite. **South Africa** Found in most southern African copper deposits. NC: Occurs in the Okiep copper district and at the Broken Hill mine. LIM: Specimens found at Phalaborwa and Musina. MP: Common at the Stavoren tin mines (where it forms green stains and coatings on fluorite). GAU: Has been found at the Willows, Kwaggafontein, Leeuwenkloof, Vergenoeg, Argent and Albert mines. **Namibia** Associated with hundreds of copper deposits throughout Namibia, notably at the Tsumeb mine (as velvet-like mats of tiny fibrous crystals, round balls, and most famously as spectacular malachite pseudomorphs after azurite). Large blocky malachite crystals have been found at the Onganja copper mine and acicular crystals occur in northern Kaokoland (with dioptase, plancheite and chrysocolla).

Balls of velvet-like malachite on dolomite, 3.5 cm. Tsumeb mine, Namibia.

Botswana Scattered copper and malachite showings are reported from the west of Maun. **Zimbabwe** Found associated with many copper deposits in several districts around Zimbabwe. **Swaziland** Occurs at Hluti, southwest of Mabasa Hill, and at Kubuta, on a ridge between the Sibowe and Lubuya streams.

Malachite on calcite, 3.8 cm. Onganja mine, Namibia.

Marcasite

FeS$_2$

Composition	sulphide
Crystal system	orthorhombic
Hardness	6–6.5
Specific gravity	4.92
Streak	green-black
Lustre	metallic

■ **Description** A metallic brass colour is characteristic of marcasite. It forms spear-shaped crystals, sometimes in groups several centimetres in diameter. Chemically, marcasite is relatively unstable and can oxidize into an efflorescent powdery white sulphate. Marcasite and pyrite are polymorphs, i.e. they are chemically identical, but pyrite is cubic, while marcasite forms orthorhombic crystals.

■ **Uses** Marcasite used to be a popular material for use in jewellery.

■ **Occurrence** **South Africa** NC: Found at the Kalahari manganese field. LIM: Reported at the Phalaborwa Complex. MP/GAU/NW/FS: Occurs in carbonaceous shales and coal seams and has been reported from several economic orebodies, e.g. the Witwatersrand and the Barberton goldfields. KZN: Has been collected from sediments at Richards Bay (where marcasite and pyrite concretions were exposed during dredging operations). **Namibia** Beautiful marcasite crystals are found at Rosh Pinah mine, southern Namibia – possibly the finest marcasite specimens from southern Africa. **Zimbabwe** Marcasite is found at the Red Wing claims, Hurungwe district, and the Mphoengs pyrite deposit, Bulilimamangwe district, where it occurs with chalcopyrite, pyrite and pyrrhotite.

Marcasite, 8.5 cm. Rosh Pinah mine, Namibia.

Microcline

KAlSi$_3$O$_8$

Composition	silicate
Crystal system	triclinic
Hardness	6–6.5
Specific gravity	2.55–2.63
Streak	white
Lustre	vitreous, pearly

Description A potassium-bearing member of the feldspar group. Crystals are white to cream, but a variety of microcline called amazonite is green. It is a widespread and common mineral of acidic plutonic rocks, such as pegmatites and granite, as well as syenite. Can occur as very large crystals and masses in pegmatites.

Uses Microcline is used in the ceramic and paint industries.

Occurrence **South Africa** NC: Found in the Pella and Pofadder areas. LIM: Amazonite occurs in the Mokopane district. **Namibia** The distribution of microcline is similar to cassiterite (tin) and ferberite-scheelite (tungsten). Common in many pegmatites in the Karibib district. Amazonite occurs south of Otjiwarongo in the Maltahöhe district, at Klein Spitzkoppe and the Erongo mountains. **Botswana** It occurs in granites and pegmatites, e.g. the pegmatite at Bodiakhudu (crystals up to 15 cm have been found), and west of Francistown.
Zimbabwe Common at the Benson pegmatites. Blue-green semi-precious amazonite comes from pegmatites in the northeast Zambezi Belt and those at Mwami, in the Hurungwe district, where the Dumbgwe, Kataha, Mago, Zunji and Ganyanhewe deposits were mined. **Swaziland** A common species in the ancient granites of Swaziland.
Mozambique Amazonite comes from the border area between southern Mozambique and Zimbabwe.

Microcline, variety amazonite, 12.5 cm. Pofadder, South Africa.

Microcline feldspar with quartz, 5.4 cm. Karibib district, Namibia.

Microcline, variety amazonite, 4.4 cm. Klein Spitzkoppe, Namibia.

Mimetite

$Pb_5(AsO_4)_3Cl$

Composition	arsenate
Crystal system	hexagonal
Hardness	3.5–4
Specific gravity	7.28
Streak	white
Lustre	resinous to subadamantine

Description A rare mineral, belonging to the apatite group. Forms hexagonal prismatic yellow to orange crystals. Mimetite is a secondary mineral found in some lead deposits.

Occurrence **South Africa** Not very common. MP: Small crystals are found at the Stavoren tin mines. GAU: Occurs at the Argent mine and the Edendale lead mine. **Namibia** Superb gem-quality mimetite comes from the Tsumeb mine, Otavi mountainland (either transparent hexagonal crystals or opaque stellate clusters). It has also been reported from the Abenab West mine. **Zimbabwe** Occurs in the Mutare area and at the Copper King mine.

Mimetite, 4.6 cm.
Tsumeb mine, Namibia.

Scattered mimetite crystals on a dark brown dolomite matrix, 9 cm. Tsumeb mine, Namibia.

A group of radiating mimetite crystals, 6 cm. Tsumeb mine, Namibia.

Molybdenite
MoS$_2$

Composition	sulphide
Crystal system	hexagonal
Hardness	1–1.5
Specific gravity	4.62–5
Streak	greenish
Lustre	metallic

Description Easy to identify as it forms metallic silver-grey crystals that are flat and hexagonal. Very soft and sectile (i.e. can bend without breaking). Molybdenite tends to soil the fingers when handled. May be confused with graphite crystals, except that graphite has a black streak. Most molybdenite is recovered from low-temperature, low-grade hydrothermal deposits.

Uses Molybdenite is the only molybdenum-bearing mineral of any economic importance. Molybdenum is used as an alloying agent and as a refractory material.

Occurrence **South Africa** NC: Occurs in the Richtersveld, west of Vioolsdrif. MP: Was mined at the Stavoren tin mines. GAU: Found in the Bushveld granites at Houtenbek, northeast of Pretoria. WC: Present in sandstones located south and southeast of Beaufort West. **Namibia** Reported from several pegmatites and porphyry copper deposits. Impressive crystals came from the Onganja copper mine. Other noteworthy deposits are at the Natas mine, Haib copper deposit, Lorelei, southeast of Rosh Pinah, Dawib Ost 61, Navachab and the Krantzberg mine. **Zimbabwe** Molybdenite occurs at the Lazeno deposit, Chipinge district, in a brecciated syenite. Attractive crystals are found at the Molly molybdenite deposit, in the Chipinge district (in quartz veins in granite), and in the Makonde and Rushinga districts. **Swaziland** It is disseminated in some granites and pegmatites, e.g. along the Komati River near the old Piggs Peak-Mbabane road.

Molybdenite on quartz, 6 cm. Houtenbek, South Africa.

Molybdenite, 8.5 cm. Onganja mine, Namibia.

Muscovite

KAl$_2$(Si$_3$Al)O$_{10}$(OH,F)$_2$

Composition	silicate
Crystal system	monoclinic
Hardness	2.5–4
Specific gravity	2.77–2.88
Streak	white
Lustre	pearly, vitreous

Description A member of the mica group of minerals, muscovite is called 'white mica' because of its silver-white sheen. It is a common rock-forming mineral in felsic igneous rocks, some sandstones and shales, and in granite and some schists. In granitic pegmatites it occurs as large crystal sheets – over a metre in diameter. There are two varieties of muscovite: sericite and fuchsite. Sericite is very fine grained, while fuchsite is the bright green chromium-rich variety.

Muscovite, 5.9 cm. Erongo mountains, Namibia.

Uses Used in electrical apparatus and by the electronics industry, and as an additive in paint, plaster and wall coatings.

Occurrence **South Africa** LIM: Sheets of over a metre are mined in the Letaba district. In the Mica district, large mica crystals occur in coarsely crystalline quartz and feldspar, often associated with garnet and apatite-(CaF). Fine-grained sericite is common at Musina, while green fuchsite is common in the ancient greenstone belt rocks. **Namibia** Ubiquitous in most pegmatites in the Brandberg-Erongo-Uis-Karibib region and Tantalite Valley. **Botswana** Common north and south of Francistown, and in the Vukwe area. Also found in granite-gneiss south/southeast of Mochudi. **Zimbabwe** Mined extensively from the pegmatites in the Mwami area, northeast of Karoi. Widespread in the Hurungwe district pegmatites. **Swaziland** Common in pegmatites, e.g. near Gege, south of Nahlozane Gorge.

Muscovite, 2.6 cm. Erongo mountains, Namibia.

Muscovite, 4.8 cm. Erongo mountains, Namibia.

Natrolite

$Na_2Al_2Si_3O_{10} \cdot 2H_2O$

Composition	silicate
Crystal system	orthorhombic
Hardness	5–5.5
Specific gravity	2.2–2.26
Streak	white
Lustre	vitreous, pearly

■ **Description** A member of the zeolite group of minerals, natrolite is always white to colourless, forming characteristic radiating sprays of acicular or thin prismatic crystals. May also be found as individual matchstick-like crystals.

■ **Occurrence** **South Africa** NC: Natrolite crystals have been produced from the Kalahari manganese field (together with apophyllite-(KF), thomsonite and inesite). In some kimberlites, colourless, transparent or opaque prismatic crystals of up to 2 cm occur. LIM/MP: Beautiful but rare crystals have been collected from some chrome and platinum mines in the Bushveld Complex. KZN: Radiating crystal sprays in amygdales occur in the Drakensburg basalts. **Namibia** Natrolite is found in basalts at the Hardap Dam, Keetmanshoop district (as crystals up to 10 cm, associated with other zeolites such as mesolite and analcime). Also occurs in cavities in the Aris phonolite south of Windhoek. **Zimbabwe** Amygdales in the Karoo basalts in western Zimbabwe contain natrolite.

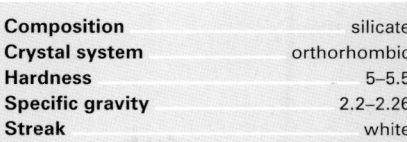

Natrolite with pink inesite, 8 cm. Wessels mine, South Africa.

Natrolite, 4 cm. Welkom goldfield, South Africa.

Orthoclase

KAlSi$_3$O$_8$

Composition	silicate
Crystal system	monoclinic
Hardness	6–6.5
Specific gravity	2.55–2.63
Streak	white
Lustre	vitreous, pearly

Description A potassium-bearing feldspar with a similar composition to that of microcline, but it crystallizes in the monoclinic system. Commonly forms large, well-shaped, tabular orange-red crystals in pegmatites. It is associated with quartz, mica and related alkali feldspars, microcline and albite. Moonstone is largely formed from orthoclase.

Uses It is used in glass, ceramics and scouring powder.

Occurrence A common rock-forming mineral found in felsic igneous rocks, syenites and some sandstones, most notably arkose, which has a high feldspar content. Orthoclase is responsible for the distinctive red-orange colour of arkoses. **South Africa** Many granites, syenites and gneisses contain orthoclase. Well-formed crystals (associated with quartz) occur in pegmatites, in the Bushveld Complex granites and at the Okiep copper mines. **Namibia** A common rock-forming feldspar found in many of Namibia's granites and gneisses. Often found as attractive well-formed crystals in pegmatites, where it is associated with other common pegmatite minerals such as schorl, quartz, microcline, topaz and tourmaline. **Botswana** Found in syenite west of Francistown. The distribution pattern of orthoclase is similar to that of microcline and it can be found in pegmatites and related igneous rocks (see also 'unakite' under epidote). **Zimbabwe** Occurs in pegmatites in most parts of the country's granitic terrain. Orthoclase is present in the rocks of the Shawa and Dorowa carbonatites.

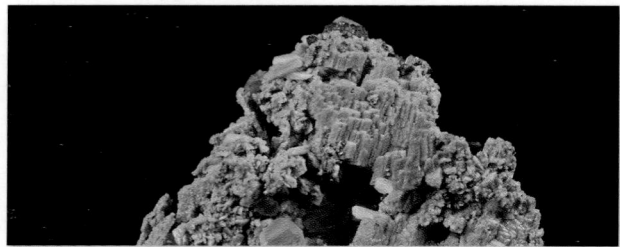

Pink orthoclase feldspar with calcite and chalcopyrite, 4.5 cm. Nababeep West mine, South Africa.

Prehnite

$Ca_2Al_2Si_3O_{10}(OH)_2$

Composition	silicate
Crystal system	orthorhombic
Hardness	6–6.5
Specific gravity	2.9–2.95
Streak	white
Lustre	vitreous, pearly

Description Occurs as beautiful, vibrant, spherical, light green to apple-green crystal aggregates in certain lavas such as basalts. This was the first mineral ever to be named after a person, i.e. its discoverer, Colonel Hendrik von Prehn. It was originally found in the Karoo dolerites of the Cradock district, Eastern Cape, South Africa.

Occurrence **South Africa** NC: Epimorphs shaped like 'angel-wing' calcite occur at Soetwater, near Calvinia (up to 50 cm long, forming interlocking plates). Unusual orange prehnite comes from the N'Chwaning II mine, Kalahari manganese field. LIM: At the Palabora mine, prehnite forms pale green crusts of compact radiating blades, and beautiful blue crystals. KZN: Similar occurrences of prehnite are found in some of the dolerite quarries, e.g. Coedmore Quarry in Durban. **Namibia** The basalts of the Goboboseb mountains, west of the Brandberg, are famous for prehnite and quartz crystals (attractive sea-green rounded aggregates and spheres, up to 10 cm diameter). Geodes lined with prehnite also occur here. The Karasburg district yields attractive smooth green botryoidal aggregates of prehnite. **Zimbabwe** Found as green botryoidal masses in some basalt vugs and occurs as a rock-forming mineral in some low-grade metamorphic rocks and igneous granites.

Prehnite, 6.7 cm. Goboboseb mountains, Namibia.

Pyrite

FeS$_2$

Composition	sulphide
Crystal system	cubic
Hardness	6–6.5
Specific gravity	5
Streak	green-black
Lustre	metallic

Description Pyrite is a characteristic brassy yellow metallic colour and is often misidentified as gold, hence its common name 'fool's gold'. It usually forms small cubic crystals, although octahedral crystals also occur. It is the most widespread and abundant sulphide and forms diagenetic concretions in many rocks.

Uses Pyrite is a source of sulphur, which is used to manufacture sulphuric acid, fertilizers, soap, detergents, matches, gunpowder, fireworks and vulcanized rubber.

Occurrence **South Africa** NC: A common sulphide in the Kalahari manganese field. LIM: Pyrite comes from the Rooiberg tin mines of the Bushveld Complex and the Murchison Range as crystals up to 2 cm. MP: Specimens recorded from the Pilgrim's Rest goldfields. Pyrite is common in coal seams. GAU: Dispersed throughout the Witwatersrand gold-bearing conglomerates as rounded 'buckshot' pyrite. **Namibia** Mined at the Otjihase, Matchless, Gorob, Tsumeb and Rosh Pinah mines. **Botswana** Found in graphite schists at Moshaneng, and in the nickel deposits in the Tati schist belt. It is common in base metal sulphide deposits and coal seams. **Zimbabwe** As in Botswana, occurs in banded-iron formations and coal seams. **Swaziland** Disseminated in gold and sulphide deposits in the Piggs Peak and Forbes Reef districts.

Tiny pyrite crystals dusting calcite, 6.1 cm.
Welkom goldfield, South Africa.

Pyrite with barite, 6.5 cm.
Otjihase mine, Namibia.

Pyrolusite

$Mn^{4+}O_2$

Composition	oxide
Crystal system	tetragonal
Hardness	2–6.5
Specific gravity	5.06
Streak	black
Lustre	metallic, dull

Description Occurs as acicular shiny black metallic crystals that may aggregate into radiating fans. More commonly pyrolusite is found in massive solid layers. It is often the mineral that forms dendrites, i.e. inorganic precipitations resembling fern-like structures and often mistakenly identified as plant fossils.

Uses Pyrolusite can be exploited for manganese, if found in sufficient concentrations.

Occurrence **South Africa** NC: Found at mines in the Kalahari and Postmasburg manganese fields. NW: Seams occur at the Ryedale mine east of Ventersdorp. KZN: Found in the Vryheid district. **Namibia** Common in the Otjosondu manganese field and the Kombat mine. Crystals occur as pseudomorphs after siderite in some of the Erongo pegmatites. Pyrolusite was common in the iron ore mined in the Otjiwarongo district. **Botswana** It is reported south of Ramotswa village and southeast of Lobatse. **Zimbabwe** It is one of the main manganese minerals and is exploited at localities in the Kwekwe and Makonde districts. Can form attractive reniform coatings and crusts. **Swaziland** Pyrolusite and manganese are found from the old Ngwenya iron mine.

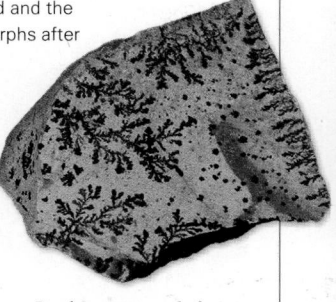

Dendrites composed of manganese oxides, 4.5 cm. Griquatown, South Africa.

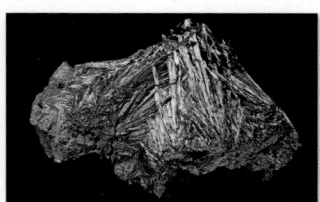

Pyrolusite crystals, 4.3 cm. Beeshoek mine, South Africa.

Dendrites are not plant fossils but inorganic formations, 6 cm. South Africa.

Pyrope
$Mg_3Al_2(SiO_4)_3$

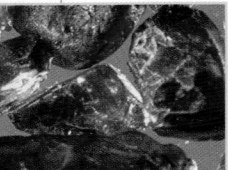

Composition	silicate
Crystal system	cubic
Hardness	7–7.5
Specific gravity	3.5–3.8
Streak	white
Lustre	vitreous

Description A member of the garnet group, pyrope forms attractive dodecahedral crystals that may be pink, pink-red, orange-red or purple-red. Rhodolite is an iron-rich red-violet variety.

Uses Rhodolite is used as a gemstone.

Occurrence **South Africa** NC/NW/FS: Commonly associated with diamond-bearing kimberlite pipes and is therefore found in alluvial gravels from diamond diggings, specifically at the Orange and Vaal rivers. **Namibia** Pyrope is found in the alluvial gravels and kimberlites of the Gibeon-Brukkaros area. **Zimbabwe** Occurs in the Selukwe and Colossus kimberlite pipes. Gem-quality rhodolite is exploited in the Mutoko and Beitbridge areas, and in the Bubi district. The Hurungwe district is another source of gem rhodolite.

Pyrope garnet variety rhodolite, 6.5 cm. Zimbabwe.

Pyroxene group – see aegirine and diopside.

Pyrrhotite

$Fe_{1-x}S$

Composition	sulphide
Crystal system	orthorhombic
Hardness	3.5–4.5
Specific gravity	4.53–4.77
Streak	grey-black
Lustre	metallic bronze-yellow to red

Description Usually forms flat platy crystals with a brassy metallic colour and lustre and may be confused with pyrite. Pyrrhotite tends to tarnish to an iridescent red-brown. It is relatively common in mafic and ultramafic rocks, associated with other sulphides such as aresenopyrite and pyrite.

Occurrence South Africa Found in the Bushveld Complex. NC: Pyrrhotite deposits occur at Okiep and Aggeneys. LIM: Found in some metal ore deposits, e.g. at Bon Accord, Musina. MP: Nkomati Nickel produces the mineral. GAU: Deposits occur in the Witwatersrand goldfields. The best pyrrhotite crystals in southern Africa came from the Mponeng gold mine, Carletonville (tabular crystals up to 5 cm in diameter, associated with galena, sphalerite, quartz and barite). **Namibia** Found 55 km southwest of Rehoboth at the Kobos copper mine, at the Matchless mine in the Otjiwarongo district, and at the Rosh Pinah mine. The Navachab gold mine and the Namib Lead mine also contained pyrrhotite. **Botswana** In the Selebi-Phikwe district, gneiss and amphibolite contain pyrrhotite with other sulphides. Small deposits are scattered through the region. **Zimbabwe** Pyrrhotite occurrence is similar to that of pentlandite, e.g they are found together at the Empress nickel mine, Trojan mine, and others.
Swaziland Occurs with chalcopyrite in Usushwana gabbro at Mhlanbanyati. Reported from the Forbes Reef area, in biotite granite in the vicinity of the waterfall on the Malolotsha River.

Pyrrhotite crystals, 3.8 cm. Mponeng mine, South Africa.

Quartz

SiO$_2$

Composition	silicate
Crystal system	hexagonal
Hardness	7
Specific gravity	2.66
Streak	white
Lustre	vitreous

Description Quartz is one of the most common minerals on Earth. It is rock-forming and is found in many igneous, sedimentary and metamorphic rocks, e.g. sandstone, granite, pegmatites, gneisses and schists – often in the form of veins. Hexagonal crystals are common, but it has many different habits including finely banded layers in agate, opal and tiger's eye. The crystals display a diagnostic conchoidal (shell-like) fracture pattern. Quartz occurs in a wide range of colours and some varieties are exploited as gemstones, e.g. amethyst, citrine, tiger's eye and carnelian.

A variety of quartz crystals, field of view 25 cm. Includes smoky quartz coated with goethite (far left), amethyst (back centre and front right) and citrine (back right and front left). Boekenhouthoek, South Africa.

Uses Quartz is the primary source of silica and silicon: silica is used in building sand and glass, while silicon chips are widely used in the electronics and computer industries.

Rock crystal quartz, 7 cm. Madagascar.

Transparent rock crystal quartz, 5.5 cm. Kopanang mine, South Africa.

Quartz varieties	
Variety	**Colour/Features**
Agate	multicoloured
▪ Blue-lace agate	powder-blue
▪ Moss agate	enclosed dendrites
▪ Picture agate	inclusions
Amethyst	purple
Aventurine	green
Carnelian	orange-red
Chalcedony	multicoloured
▪ Mtorolite	green
Chrysoprase	apple-green
Citrine	yellow
Hyalite	white, yellow
Jasper	red
▪ Brecciated	red, white
Milky quartz/rock crystal	white/colourless, transparent
Rose quartz	pink
Smoky quartz	grey, black, transparent
Tiger's eye, Pietersite	striated brown, blue

Quartz with clay inclusions, 9.8 cm. Goboboseb mountains, Namibia.

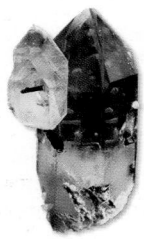

Smoky quartz with transparent rock crystal quartz and schorl tourmaline, 10.2 cm. Mozambique.

MINERALS **89**

QUARTZ VARIETY **MILKY/ ROCK CRYSTAL QUARTZ**

Description As its name suggests, milky quartz is opaque white. It does not generally form well-shaped crystals, but instead occurs as masses or lumps, commonly filling quartz veins in other rocks such as granite. Rock crystal quartz is transparent and colourless, often water-clear. It is more common as well-formed, aesthetic hexagonal crystals.

Occurrence The two most common varieties of quartz.
South Africa NC: It is widespread and common in Namaqualand. Thousands of quartz crystals were found at Jan Coetzee mine. Milky/ rock crystal quartz also comes from the Springbok-Steinkopf-Onseepkans areas (where some crystals are known as 'Herkimer diamonds'). MP/GAU: Milky/rock quartz comes from KwaNdebele, between Pretoria and Groblersdal (see also amethyst and citrine). Transparent quartz crystals and crystal groups come from the Witwatersrand gold mines.

Smoky quartz on milky quartz, 12.8 cm. Steinkopf district, South Africa.

Namibia It occurs in many of the pegmatites in the Karibib, Omaruru and Swakopmund districts. It is also found at Klein Spitzkoppe and Erongo, in the Gamsberg region and in the basalts at Tafelkop, west of the Brandberg. In southern Namibia pegmatites have produced milky/rock crystal quartz and amethyst.
Botswana Rock crystal quartz comes from the Selebi-Phikwe area, while many quartz veins are found in the Tati district, and north of Francistown. **Zimbabwe** Quartz crystals are common in pegmatites in the country. **Swaziland** Quartz is widespread in Swaziland's pegmatites. Crystals up to 35 cm are found in the Piggs Peak district.

QUARTZ VARIETY **AGATE**

Description Agate is characterized by distinct layering or banding formed by multiple thin laminations of chalcedony (a quartz variety in which the crystals are invisible to the naked eye). Many agates are found filling amygdales (gas cavities) in basalts and other volcanic rocks. The variety in which agate encloses delicate leaf-like inclusions of secondary minerals is called moss agate.

■ Occurrence **South Africa** NC/LIM: Agate
occurs in many alluvial diamond-digging
dumps and the alluvial sediment of the
Orange and Limpopo rivers.
GAU: Occurs in the alluvial sediment
of the Vaal River. NW: The
Lichtenburg diamond diggings
contain moss agate. KZN: The
Drakensberg basalts contain agate.
Blue-grey agates up to a metre in
diameter are found in the Jozini-
Pongolapoort area. WC: Agate also
occurs in the alluvial sediment of the
Caledon River. EC: Found in basalts.

Quartz, variety agate, 6.8 cm. Lesotho.

Namibia Plentiful on some beaches,
particularly Agate Beach, Lüderitz. In
southern Namibia, southwest of
Karasburg, a famous blue-lace agate
deposit is found in veins associated
with dolerite. North of the Ugab
River brown, grey, red and yellow

Quartz, variety agate, 8.4 cm.
Zimbabwe.

alluvial agates are eroded out from basalt. The diamond-bearing gravels
also contain abundant agates. **Botswana** Well-known pink and cream
agates come from alluvial deposits in the Tuli region, several kilometres
north of Pontdrif. **Zimbabwe** Agate is found filling cavities (amygdales
and geodes) in basalts, most notably in the Tsholotsho, Bumi Hills and
Featherstone areas. **Swaziland** Agates come from the southern region,
in the Lebombo basalt and rhyolites. **Mozambique** Found near the
southern Mozambique/Zimbabwe border. **Lesotho** Occurs in the
Drakensberg basalts, filling amygdales in many of the streams. Fine
yellow and blue-banded examples come from the upper Sani Pass region.

Quartz, variety agate, lining a geode, 15.5 cm. South Africa.

QUARTZ VARIETY **AMETHYST**

■ Description Amethyst is a mauve to purple variety of quartz. Ions of Fe^{3+} in the atomic lattice of the quartz give it its colour. Crystals can sometimes be found composed of a combination of amethyst, smoky quartz and clear rock crystal. These specimens are very beautifully colour-zoned.

Quartz, variety amethyst, 11.2 cm. Boekenhouthoek, South Africa.

■ Occurrence Many southern African localities. **South Africa** NC: Occurs south of Pofadder, also in the Keimoes, Kakamas and Augrabies districts, and along the Orange River. **MP**: KwaNdebele, northeast of Pretoria, has some of the finest South African amethyst. **KZN**: Beautiful amethyst-lined geodes occur in the Lebombo rhyolites. **Namibia** High-quality crystals come from basalts at Tafelkop, west of the Brandberg, and crystals line geodes at Sarusas on the Skeleton coast. Also known in the Grootfontein district in northern Namibia (in quartz-calcite veins hosted in marble and limestone). **Zimbabwe** Found in basalts in the Tsholotsho and Hwange districts, and in the pegmatites of the Hurungwe region. Also Manzinyama in the Nyamandhlovu district, the Coronet and Pat pegmatites in the Hurungwe district, the pegmatites and basalts in the Beitbridge district (Featherstone and Mutoko areas), and the Bumi district.

Quartz, variety amethyst, 6 cm. Goboboseb Mountains, Namibia.

QUARTZ VARIETY **AVENTURINE**

■ Description Aventurine is an apple-green variety of chalcedony. Minute flecks of chrome-bearing muscovite (fuchsite) that are included in the mineral give it a sparkling appearance.

■ Occurrence **South Africa** LIM: Found on the farm Santor, in the Soutpansberg and in the Gravelotte and Leydsdorp districts. **MP**: Also occurs in the Barberton district. **Zimbabwe** Deposits found in the northern Beitbridge district, between Gwanda and Beitbridge and at Jopempe (as veins of aventurine 3–15 m wide). High-quality aventurine comes from the Altitude mine, Masvingo district.

Quartz, variety aventurine, 5.5 cm. Zimbabwe.

Description Carnelian is an orange-brown variety of chalcedony.

Occurrence **South Africa** NC: It occurs in the Hay and Prieska districts and in the alluvial diamond gravels around Kimberley, Barkly West and Lichtenburg. LIM: Lapidary quality carnelian is found near Musina. KZN: Carnelian occurs in Lebombo volcanics. **Namibia** Carnelian occurs along the Skeleton Coast, particularly at Sarusas. **Zimbabwe** It is found associated with basalt, as layers or bands in amygdales and vugs, or as alluvial specimens weathered out from basalt. **Swaziland** It is found in the southeast, in Lebombo rhyolites and associated volcanic rocks.

Quartz, variety carnelian, 5.1 cm. Sarusas, Namibia.

QUARTZ VARIETY **CHALCEDONY**

Description Chalcedony is micro-crystalline quartz, usually white, grey, black or pale blue in colour. The attractive green variety of chalcedony is known as mtorolite.

Occurrence **South Africa** NC/NW/FS: Chalcedony is common in the alluvial gravels of the Vaal and Orange rivers. KZN: It occurs in the Drakensberg basalts and rhyolites. **Namibia** High quality, jewellery-grade, semitransparent blue chalcedony occurs about 150 km northeast of Okahandja. It is also found near Rössing Siding (as yellow, cream and blue stalactites and botryoidal layers in cavities in jasper). **Zimbabwe** Chalcedony is common in amygdales and geodes in basalts. Mtorolite occurs in veins in the rocks of the Great Dyke. **Swaziland** Green chalcedony occurs in the northeastern region.

Quartz, variety chalcedony, 6.8 cm. Rössing Mountain, Namibia.

Quartz, variety chalcedony, 4.6 cm. N'Chwaning I mine, South Africa.

Quartz, variety chalcedony, 2.1 cm. Palabora mine, South Africa.

■ **Description** An attractive apple-green variety of quartz, chrysoprase gets its appealing colour from trace amounts of nickel.

■ **Occurrence Namibia**
Chrysoprase is found near Rehoboth and in the Karasburg district. It is also known from the Kaokoland, 45 km northwest of Otjovazandu. **Zimbabwe** Good quality chrysoprase comes from the eastern regions of the country.

Quartz, variety chrysoprase, 6.4 cm. Zimbabwe.

■ **Description** Citrine is the transparent yellow variety of quartz. It can be faceted and used to make jewellery.

■ **Occurrence South Africa** MP: Citrine occurs at KwaNdebele. FS: It is also found at Jagersfontein. **Namibia** Occurs in tantalite-bearing pegmatites south of the Rubikon mine, on the farms Okongava Ost 72 and Otjimbingwe 104. **Zimbabwe** Granites in the Harare South region and due west and north of Marondera have veins containing citrine.

Quartz, variety citrine, 6.1 cm. Zambia.

Faceted 65-carat citrine from Marondera, Zimbabwe.

Quartz, variety citrine, 5.5 cm. Zambia.

Description Hyalite is an opalline variety of quartz. It forms botryoidal coatings and masses, often on other minerals. It also occurs as tiny, smooth, glass-like botryoidal beads, aggregated together.

Occurrence Namibia
Yellow and yellow-green botryoidal opalline silica (hyalite) is found coating feldspar, schorl tourmaline, quartz and aquamarine in pegmatites in the Erongo mountains. It fluoresces an intense yellow-green under ultraviolet light. Hyalite is also found in some pegmatites at Klein Spitzkoppe.

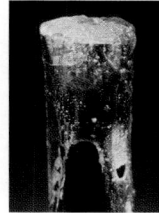

Glassy hyalite coating a schorl tourmaline, 3.3 cm. Erongo mountains, Namibia.

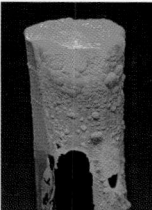

The same specimen, but fluorescing under ultraviolet light.

Description Jasper is bright red iron-rich chert and can be found in different forms – from solid red varieties to an attractive mosaic of broken jasper fragments cemented by white quartz, which is known as brecciated jasper. Another form consists of banded red jasper intermixed with yellow-orange chert and layers of silver hematite ore. 'Bloodstone' contains green chert spotted with inclusions of red jasper.

Occurrence South Africa NC: Large commercial deposits occur near Griquatown. MP/KZN: Jasper is found in some very ancient Archaean rocks, e.g. in the Barberton and Pongola districts. **Namibia** Found northwest of Rössing Siding. Layers of jasper associated with banded-iron formations occur in the Otavi mountainland and in Kaokoland. **Botswana** Found in banded-iron formations located in the southern areas of the country, south of Khakhea. Layers of jasper occur in the Tati schist belt in the eastern border region. It is also found at Matsijole, southeast of Francistown. **Zimbabwe** Common in outcrops of banded-iron formation rocks, interlayered with hematite and chert. **Swaziland** In the eastern parts of the country 'Koi' jasper occurs (where the mineral is interlayered with semitranslucent chert).

Quartz, variety jasper, 9.8 cm. Griquatown district, South Africa.

QUARTZ VARIETY **ROSE QUARTZ**

■ **Description** The pink colour of rose quartz is caused by minute traces of titanium. Rose quartz crystals are extremely rare.

■ **Occurrence** **South Africa** NC: Found in veins and pegmatites near Kenhardt and Keimoes and at Riemvasmaak. Rose quartz is also present in the Goodhouse-Wolftoon area and the Orange River Valley. LIM: Found in the granites of the Soutpansberg district, and at Selatidrift near Gravelotte. **Namibia** Rose quartz comes from Rössing Siding, east of Swakopmund. It occurs in pegmatites in southern Damaraland and the Karibib and Karasburg districts. The pegmatites south of Uis and east of Cape Cross contain translucent rose quartz. **Zimbabwe** Good quality rose quartz comes from pegmatites in the Hurungwe and Mutoko districts and the Beitbridge region.

Rose quartz, 17.2 cm. Northern Cape, South Africa.

QUARTZ VARIETY **SMOKY QUARTZ**

■ **Description** Smoky quartz is typically transparent and light grey, grey-brown, dark grey or black. The grey-brown-black colour could be due to radiation from natural sources or it may result from minute inclusions of other minerals. Alternatively, smoky quartz may owe its colour to the presence of trace amounts of aluminium ions, Al^{3+}.

■ **Occurrence** **South Africa** NC: Good quality specimens come from pegmatites in the province. The Jan Coetzee mine, Springbok district, produced spectacular large crystals. WC: Also occurs in many granitic areas such as on Paarl Mountain. **Namibia** It is found in cavities in basalts at Tafelkop, west

Smoky quartz, 7.5 cm. Goboboseb mountains, Namibia.

Smoky quartz, 9.5 cm. Zomba Mountain, Malawi.

of the Brandberg. Occurs in pegmatites in the Gamsberg region and the Karibib, Usakos and Swakopmund districts, e.g. at Neu Schwaben (as large, sceptred smoky quartz crystals). Beautiful transparent dark grey to black smoky quartz comes from Klein Spitzkoppe and the Erongo mountains. **Zimbabwe** Smoky quartz occurs in granites south of Harare and the Marondera region. It is also found in the Hurungwe district pegmatites.

QUARTZ VARIETY **TIGER'S EYE**

Description Tiger's eye forms from the silicification of weathered asbestos – hydrothermal fluids containing dissolved silica percolate through the asbestos and the silica then replaces and recrystallizes the asbestos as tiger's eye. The colours of tiger's eye depend on the degree and intensity of weathering. It is typically yellow-brown, but can also occur as grey-green and yellow-green varieties. The blue variety forms by silicification of unweathered blue crocidolite, while red tiger's eye is produced artificially by heating yellow-brown tiger's eye to a temperature of 400 °C.

Quartz, variety tiger's eye, 9.4 cm. Griquatown district, South Africa.

Occurrence **South Africa** NC Tiger's eye outcrops in the Griqualand West region, where it is commercially mined. **Namibia** Pietersite, a variety of tiger's eye, occurs northeast of Outjo, formed by the brecciation and silicification of blue crocidolite (i.e. crocidolite fragments are 'cemented' together by silica).

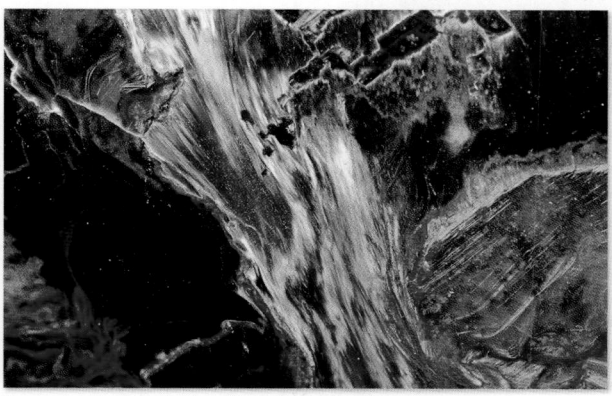

Quartz, variety pietersite, 6 cm. Outjo district, Namibia.

Rhodochrosite

$Mn^{2+}CO_3$

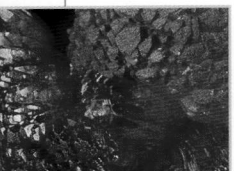

Composition	carbonate
Crystal system	hexagonal
Hardness	3.5–4
Specific gravity	3.7
Streak	white
Lustre	vitreous, pearly

Description Rhodochrosite is a relatively rare mineral worldwide, but is highly sought-after by mineral collectors. It is a manganese carbonate and forms from the oxidation of manganese ore. The crystals are rhombohedral or scalenohedral ('dog's tooth'-shaped), but the mineral can also occur as either rounded or 'wheat-sheaf' aggregates. A red or pink colour is diagnostic, but it can also be pale red to brown-red.

Rhodochrosite, 4.5 cm. N'Chwaning II mine, South Africa.

Occurrence **South Africa** NC World-famous blood-red crystals came from the Kalahari manganese field. Varieties included dark red balls and intense dark red scalenohedrons of up to 7 cm in length on a jet-black manganite matrix. Transparent fragments are faceted into gemstones. Small opaque pink rhombohedra came from the adjacent Wessels mine. It has also been found at the Broken Hill mine, Aggeneys.

Namibia Rhodochrosite has been reported from the Ai-Ais lead mine (associated with galena and malachite) and has also been found at the Kombat mine and, very rarely, in the Erongo mountains. **Zimbabwe** Has been found at the Iona mine, Mutare district, and the Kingsley mine, Bindura district.

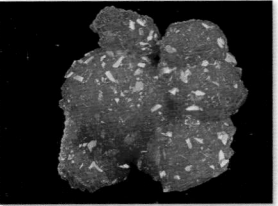

Cluster of 'dog's tooth' scalenohedral rhodochrosite, 3.2 cm. N'Chwaning I mine, South Africa.

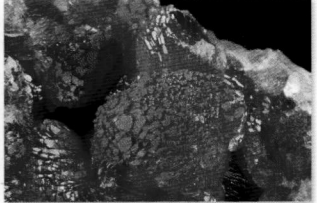

Highly lustrous rhodochrosite, 3.6 cm. N'Chwaning II mine, South Africa.

Bundles of 'wheat sheaf' rhodochrosite, 4.2 cm. Hotazel mine, South Africa.

Riebeckite

$Na_2(Fe^{2+},Mg)_3Fe_2^{3+}Si_8O_{22}(OH)_2$

Composition	silicate
Crystal system	monoclinic
Hardness	5
Specific gravity	3.32–3.38
Streak	blue-grey
Lustre	vitreous, silky

Description In southern Africa, riebeckite is most common as bright blue crocidolite asbestos. When crocidolite weathers, it becomes brown and brittle. If silicification of these weathered fibres then takes place, tiger's eye is formed.

Occurrence **South Africa** This asbestiform amphibole species is found in banded iron formations from Pomfret to Prieska. **Namibia** Occurs on the farm Groot Aub 267 in the Rehoboth district. A silicified variety, 'pietersite', is used as a gemstone (see tiger's eye).

Layers of riebeckite asbestos, 11 cm. North West, South Africa.

Riebeckite asbestos, 10.5 cm. South Africa.

Rutile

TiO$_2$

Composition	oxide
Crystal system	tetragonal
Hardness	6–6.5
Specific gravity	4.23
Streak	pale brown to yellow
Lustre	adamantine, submetallic

■ **Description** Rutile forms elongate acicular crystals or more robust prismatic crystals that are often twinned. It is typically dark brown to dark red and can be highly lustrous. Rutile is an accessory mineral in some igneous and metamorphic rocks. As a titanium oxide it is resistant to weathering and breakdown, hence concentrations occur in quartz sand on coastal beaches with other 'heavy mineral' species such as ilmenite, zircon, garnet and monazite.

Rutile included in quartz, 4.5 cm. Zambia.

■ **Uses** Titanium has many applications. An important use is in manufacturing the lightweight alloys required by the aerospace industry.

■ **Occurrence South Africa** LIM: Crystals are found in the volcanic tuffs at the Pilanesberg Complex and at the Glenover phosphate mine. KZN: Rutile is mined at Richards Bay and at Umgababa. **Namibia** Rutilated quartz comes from the farm Rooisand in the Gamsberg region, while Onganja copper mine is famous for red rutile crystals. Large euhedral crystals occur at Giftkuppe in the Omaruru district, where rutile was commercially mined from veins of granite. **Botswana** Minor amounts of rutile are found in the Halfway Kop kyanite deposit. **Zimbabwe** Occurs in the Mwami pegmatite region. **Swaziland** The mineral is found in schists at Mhlabane Mountain, Gege district, with magnetite and zircon. It is also reported from a biotite gneiss near the Malolotha River waterfall in the Forbes Reef area. **Mozambique** Economic deposits of rutile and associated heavy minerals occur along the southern coastline.

Rutile crystal, 1.4 cm. Windhoek district, Namibia.

Scheelite

CaWO$_4$

Composition	tungstate
Crystal system	tetragonal
Hardness	4.5–5
Specific gravity	6.1
Streak	white
Lustre	vitreous, adamantine

Description Scheelite crystals are typically heavy because of their tungsten content. They are commonly colourless, but can also be found as attractive yellow, orange, green or red pseudo-octahedral crystals. Scheelite fluoresces readily and this characteristic allows it to be found at night, using an ultraviolet light.

Uses Scheelite is economically important and is mined for tungsten. (See ferberite for the uses of tungsten.)

Occurrence **South Africa** NC: Scheelite is found in some pegmatite deposits in Namaqualand, notably in the Okiep district. LIM: Occurs in the Murchison mountain range. MP: Scheelite was found at the Stavoren mines in the Bushveld Complex and at Pilgrim's Rest and Barberton. **Namibia** It occurs in granites, skarns, pegmatites and some Alpine cleft deposits. Enormous scheelite crystals, weighing up to 60 kg, have been found on the farm Kos in the Gamsberg area. Green copper-bearing scheelite (cupro-scheelite) comes from the farm Natas 220, 120 km southwest of Windhoek. Also occurs in pegmatites in the Warmbad district, the Brandberg West and Krantzberg mines and in the Erongo region. **Botswana** Relatively widespread in the Tati schist belt, usually in quartz veins. **Zimbabwe** A common species previously exploited at several hundred deposits, e.g. in the Bulawayo area, in north and northeast Zimbabwe and elsewhere. **Swaziland** Found in tin-bearing gravels near Sinceni and in the Mbabane district. At Forbes Reef, on the Malolotsha River, scheelite occurs with calcite, albite, rutile and pyrrhotite in biotite gneiss.

Scheelite, 3 cm. Northern Cape, South Africa.

Serpentine – see serpentinite (metamorphic rocks).

Schorl

$NaFe_3^{2+}Al_6(BO_3)_3Si_6O_{18}(OH)_4$

Composition	silicate
Crystal system	hexagonal
Hardness	7
Specific gravity	3.1–3.25
Streak	white
Lustre	vitreous, resinous

Description Schorl is the iron-rich member of the tourmaline group and the most common tourmaline species. It forms distinctive prismatic opaque jet-black crystals, but can also occur in other colours such as deep red or deep mauve. It is a common accessory mineral in pegmatites.

Occurrence **South Africa** NC: A relatively rare mineral in the province's pegmatites, but it is found sporadically in many of them. The striated black crystals may be crudely formed or euhedral, and up to 25 cm in length. LIM/MP: In the pegmatites within these provinces, schorl is commonly found embedded in quartz. It is a common accessory mineral in the defunct Bushveld tin mines, occurring as radiating stellate groups. **Namibia** Schorl from the Erongo mountain pegmatites is among the finest in the world: individual, extremely shiny, prismatic, jet-black crystals up to 40 cm long are known. Schorl is found elsewhere in Namibia, scattered in many other pegmatites. **Zimbabwe** As in Namibia and South Africa, schorl is common in pegmatites and is found in many deposits in Zimbabwe.

Radiating fan-shaped group of schorl crystals, 15.2 cm. Karibib, Namibia.

Schorl tourmaline, 14.3 cm. Erongo mountains, Namibia.

A spray of schorl crystals in mica schist, 6.2 cm. Northern Cape, South Africa.

Siderite

$Fe^{2+}CO_3$

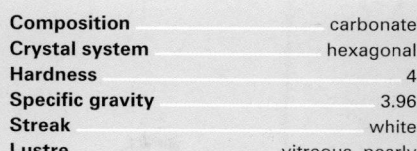

Composition	carbonate
Crystal system	hexagonal
Hardness	4
Specific gravity	3.96
Streak	white
Lustre	vitreous, pearly

Description Occurs as brown to green rhombohedral crystals and is a relatively common constituent in the matrix of some sedimentary rocks, e.g. sandstones and banded-iron formations. Sometimes forms layers and nodules in existing sedimentary rocks and can also be found in hydrothermal ore veins.

Occurrence **South Africa** Siderite crystals are rare. NC: Spectacular siderite-sphalerite pseudomorphs after calcite were found at the Broken Hill mine at Aggeneys. GAU: Siderite comes from the Argent mine. KZN: Some siderite is found at the Clairwood Quarries, and in the Durban and Vryheid districts. **Namibia** Occurs in banded-iron formations, particularly in the south and southeast of the country. Large 20 cm crystals, partly pseudomorphed to goethite (i.e. some of the original siderite still remains), have been collected from the Erongo mountains. Arsenic-rich siderite was found at the Tsumeb mine.

Siderite and quartz, 9.6 cm.
Erongo mountains, Namibia.

Zimbabwe Occurs as nodules and layers in some sedimentary rocks and in banded-iron formations. **Swaziland** Siderite came from the Ngwenya iron ore mine, and is also found as a rock-forming mineral in banded-iron formations.

Transparent siderite crystals, 5 cm.
Tsumeb mine, Namibia.

Smithsonite

$ZnCO_3$

Composition	carbonate
Crystal system	hexagonal
Hardness	4–4.5
Specific gravity	4.3–4.45
Streak	white
Lustre	vitreous, pearly

Description Smithsonite can be found as well-formed rhombohedral crystals or as botryoidal masses. The crystals resemble calcite but are heavier. It is commonly associated with zinc sulphide deposits, where the oxidation of the primary zinc ore forms smithsonite. It is white to colourless if pure, but the presence of trace elements produces a wide range of colours including yellow, green, pink, blue, brown and orange.

Occurrence South Africa Found as small crystals in lead-zinc deposits, e.g. in the Ottoshoop district and the Argent and Broken Hill mines. **Namibia** Superb colourful crystals up to 7 cm originated from the Tsumeb mine, including green and blue cupro-smithsonite, pink cobalt- and mangan-smithsonite and yellow, orange and caramel cadmium-smithsonite. White smithsonite, with willemite and cerussite, came from the Abenab and Berg Aukas mines. Smithsonite is also found at the Skorpion mine. **Swaziland** Found in a thin epidote vein 5 km east of Nkambeni hill.

Rosettes of smithsonite, 3.4 cm. Berg Aukas mine, Namibia.

'Dog's-tooth' smithsonite, 2.8 cm. Berg Aukas mine, Namibia.

Cobalt-rich smithsonite, 10 cm. Tsumeb mine, Namibia.

Green, copper-rich smithsonite, 4 cm. Tsumeb mine, Namibia.

Sodalite

$Na_8Al_6Si_6O_{24}Cl_2$

Composition	silicate
Crystal system	cubic
Hardness	5.5–6
Specific gravity	2.14–2.4
Streak	white
Lustre	vitreous, greasy

Description Sodalite is a distinctive, very dark blue colour and can be confused with lapis lazuli. It usually contains bands or layers of white ankerite, yellow cancrinite and pale pink analcime.

Uses It is used for cladding, tiles, carvings, stonework and jewellery.

Occurrence **South Africa** NW: Minor amounts are found in the Pilanesberg. **Namibia** The largest deposit in southern Africa is in Kaokoland, 10 km west of Swartbooisdrif, hosted in carbonatite and syenite. It is sporadically exploited for ornamental stonework. Beautiful dark blue masses up to 8 m are found in this deposit. **Zimbabwe** Sodalite was previously quarried in the Mwami and Hurungwe areas.

Sodalite, 4 cm, from northern Namibia.

Sodalite, 8.4 cm. Swartbooisdrif, Namibia.

Spessartine

$Mn_3^{2+}Al_2(SiO_4)_3$

Composition	silicate
Crystal system	cubic
Hardness	7–7.5
Specific gravity	3.8–4.25
Streak	white
Lustre	vitreous

Description This member of the garnet group can be very valuable – the transparent vibrant orange variety commands high prices. It is most common in the form of brown, reddish-brown or red dodecahedral crystals.

Uses Spessartine is used as a gemstone.

Occurrence **South Africa** NC: It occurs sporadically in pegmatites. LIM: Spessartine has been reported from the schists of the Murchison mountain range. **Namibia** World-class bright orange spessartine gemstones from northern Kaokoland have been dubbed 'Mandarin garnet' in the gem trade. Spessartine is also found in the Otjosondu manganese field and the Kombat mine. **Zimbabwe** It occurs in metamorphosed granite in the Hurungwe and Masvingo district, and crystals have come from the Good Days beryl-lithium pegmatite, Mutoko district. **Swaziland** Crystals of up to 1 cm occur in a pegmatite 12 km southwest of Mbabane. At Gege, spessartine is found as individual crystals or as layers with iron ore in quartzite.

Spessartine garnet, 3.6 cm. Kaokoveld, Namibia.

Sphalerite
(Zn,Fe)S

Composition	sulphide
Crystal system	cubic
Hardness	3.5–4
Specific gravity	3.9–4.1
Streak	white to pale brown
Lustre	resinous, adamantine

Description Crystals typically display a resinous waxy lustre. They are most commonly dark orange to amber (if backlit), but may also be found in shades of brown, black or, rarely, green.

Uses Sphalerite is the main ore mineral of zinc, used in bronze and brass alloys. It is also a filler in paint and rubber, an oxide in feed additives and soil rejuvenation, and is used in zinc batteries.

Occurrence Sphalerite is common in some sedimentary and hydrothermal ore deposits. **South Africa** NC: There is a deposit at Bushy Park, near Griquatown, and deposits also occur at the Gamsberg and Broken Hill mines. GAU/FS: Rare sphalerite crystals up to 5 cm occur at some Witwatersrand gold mines, in the Welkom district and Carletonville. NW: Beautiful honey-coloured crystals in dolomite come from the Pering mine at Reivilo district, southwest of Vryburg. These are found with grey dolomite crystals, galena and quartz. **Namibia** Found in most of the lead-zinc deposits in Namibia, as well as in the Otavi mountainland, Kaokoland and the Grootfontein district.

Sphalerite crystals, 3.1 cm. Pering mine, South Africa.

Zimbabwe It is commercially mined at many lead deposits, because galena and sphalerite are very often found together, e.g. in the Gokwe, Kwekwe and Nyanga districts. **Swaziland** Sphalerite crystal clusters as big as oranges are found in dolomite at the She mine at Forbes Reef. It is associated with the barite deposit near Oshoek, and is found in the unusual epidote vein outcropping east of Nkambeni hill.

Sphalerite and dolomite, 7.2 cm. Pering mine, South Africa.

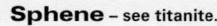

Sphene – see titanite.

Spinel
MgAl$_2$O$_4$

Composition	oxide
Crystal system	cubic
Hardness	7.5–8
Specific gravity	3.58
Streak	white
Lustre	vitreous to dull

Description Spinel is identified by its distinctive octahedral crystals. Its colour is variable, depending on trace element contamination and crystals may be black, dark blue, green, beige, pink or red.

Occurrence It is found in metamorphic serpentinites, gneisses, calc-silicates and marble, and some mafic igneous rocks. **South Africa** NC: Zinc-bearing spinel, gahnite, occurs at Aggeneys. LIM: Some kimberlites in the province contain chrome-spinel. MP: Spinel also occurs on the farms Groot Hoek 256 KT and Thorncliffe 374, in the Lydenburg district. NW: Pink-red grains of spinel are present in the diamondiferous gravels at Lichtenburg. KZN: Marbles from Marble Delta have also produced small grains of the mineral. **Namibia** Sharp, dark grey to black octahedrons of up to 2 cm are found close to Rössing, and large crystals occur in a skarn deposit on the farm Okahua, about 30 km south of Otjiwarongo.

Spinel embedded in calcite, 2 cm octahedral crystal. Rössing area, Namibia.

Staurolite

$(Fe^{+2},Mg,Zn)_2Al_9(Si,Al)_4O_{22}(OH)_2$

Composition	silicate
Crystal system	monoclinic
Hardness	7–7.5
Specific gravity	3.65–3.83
Streak	white to pale grey
Lustre	vitreous, resinous

Description A metamorphic mineral, staurolite forms prismatic dark brown to black crystals that are very often twinned. Its name, derived from the Greek word for a cross, is a reference to this shape.

Occurrence Staurolite is usually found in schists and associated metamorphic rocks. **South Africa** NC: Occurs at Aggeneys and other Namaqualand localities, e.g. west of Upington. **Namibia** Crystals up to 8 cm long (many twinned) are found in the Namib Desert Park near the Gorob mine. Staurolite often occurs in heavy-mineral sand deposits, especially where the source rock is metamorphic schist. **Zimbabwe** Staurolite schist is common at the Kondo mine. It is hosted in pegmatite and the schists of Hurungwe, Mutoko and Mount Darwin districts. **Swaziland** Reported from a locality 11 km east of Goedgegun, which is close to Ferreira's Station.

Cluster of staurolite crystals in mica schist, 10 cm. Gorob mine, Namibia.

Twinned staurolite crystals, Gorob mine, Namibia. The central twinned crystal is 3.4 cm.

■ Stibnite

Sb$_2$S$_3$

Composition	sulphide
Crystal system	orthorhombic
Hardness	2
Specific gravity	4.63–4.66
Streak	grey
Lustre	metallic, brilliant

■ **Description** The prismatic crystals are usually elongated and metallic silver-grey. Stibnite is soft and can soil the fingers.

■ **Uses** It is the main ore mineral of antimony, which is used in batteries and metal alloys.

■ **Occurrence** **South Africa** LIM: The largest deposit in southern Africa is found close to Gravelotte. This is geologically the oldest antimony deposit in the world. Most stibnite mined is massive, rather than forming euhedral crystals. MP: It is also present in the Barberton district. **Zimbabwe** Found in many deposits in Zimbabwe, mostly in the Kwekwe district (in hydrothermal quartz veins, together with gold and other sulphides). The Globe and Phoenix gold mine is world-famous for sword-like crystals of kermesite that have pseudomorphed after stibnite crystals. **Botswana** Found at a few localities in the Tati schist belt, where stibnite occurs with gold in quartz veins. **Swaziland** Acicular crystals of stibnite are found at the She mine in the Forbes Reef district, while small stibnite crystals came from the Primrose and Avalanche mines.

Stibnite crystals, 3 cm. Consolidated Murchison mine, South Africa.

Stichtite

$Mg_6Cr_2(CO_3)(OH)_{16} \cdot 4H_2O$

Composition	carbonate
Crystal system	hexagonal
Hardness	1.5–2
Specific gravity	2.16
Streak	pale-lilac to white
Lustre	pearly, waxy, greasy

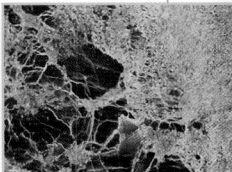

■ **Description** A member of the hydrotalcite group of minerals, stichtite is one of the few purple minerals known. Ranging from lilac to mauve in colour, it has a very soft soapy-greasy texture. It occurs in veins or lumps, not large crystals, and is associated with serpentinites.

■ **Occurrence** **South Africa** LIM: Stichtite has been recorded from the Consolidated Murchison mine, near Gravelotte. MP: Kaapsehoop in the Barberton district is famous for stichtite lumps up to 20 cm in diameter and stichtite veins hosted in green serpentine. **Zimbabwe** Stichtite is known from the Shabanie asbestos mine, which is in the Shurugwi district.

Stichtite in serpentinite, with minor white asbestos veins, 11.5 cm. Kaapsehoop, South Africa.

Polished sample of stichtite, 7.5 cm. Kaapsehoop district, South Africa.

Stilbite

$NaCa_2Al_5Si_{13}O_{36} \cdot 14H_2O$

Composition	silicate
Crystal system	monoclinic
Hardness	3.5–4
Specific gravity	2.09–2.2
Streak	white
Lustre	vitreous, pearly

Description Stilbite is a member of the zeolite group of minerals. It has a distinctive pearly lustre and often forms bowtie-shaped or fan-like crystals that are characteristically white or cream. It is found in cavities in volcanic rocks, notably basalt.

Occurrence **South Africa** LIM: Found at the Palabora mine and the Mooinooi Chrome mine. KZN: It is also found in road cuttings on the Carlisle's Hoek road, and near the top of Naude's Nek pass and Witsieshoek (with other zeolite species). EC: Occurs in the Drakensberg basalts, Barkly East district. Salmon-pink to white opaque wheat-sheaf bundles are found in the Barkly East and Rhodes districts. **Namibia** Very good crystals – opaque white bowties and wheat sheafs up to 15 cm – are found in the basalts at Grootberg Pass, 90 km east of Khorixas, on the road to Sesfontein. **Zimbabwe** Stilbite occurs southwest of Bulawayo and at the Jessie copper mine in the vicinity of Victoria Falls. Crystals of stilbite fill joints and fractures in amphibolite that outcrops in the Mazoe River, 5 km upstream from the Southern Cross tungsten mine in the Rushinga district.

Stilbite, 1.8 cm. Barkly East district, South Africa.

Stilbite on quartz, 10.5 cm. Grootberg Pass, Namibia.

Sturmanite

$Ca_6(Fe^{3+},Al,Mn^{2+})_2(SO_4)_2(B[OH]_4)(OH)_{12} \cdot 25H_2O$

Composition	calcium sulphate hydroxide
Crystal system	hexagonal
Hardness	2.5
Specific gravity	1.85
Streak	light yellow
Lustre	vitreous

Description A rare mineral that varies in colour from white to pale apricot, yellow, lemon yellow, dark orange and yellow-brown, though it does tend to discolour with time to a dirty brown. It is soft, forming brittle crystals. Larger crystals display colour zoning.

Occurrence South Africa NC: A type-locality species from the Kalahari manganese mines, i.e. it was first discovered there. It remains unique to this country yet thousands of specimens have been collected since its discovery in the early 1980s. The N'Chwaning II mine produced crystals up to 30 cm and prismatic crystals of 4 cm are relatively common. Many crystals discolour over time and partially alter to gypsum. Commonly associated minerals include andradite, barite, calcite, celestine, grossular, gypsum, hausmannite, hematite, kutnohorite, manganite and rhodochrosite.

Group of large sturmanite crystals, 11 cm. Kalahari manganese field, South Africa.

Large sturmanite crystal, 15.2 cm. N'Chwaning II mine, South Africa.

Sugilite

$K, Na_2(Fe^{2+}, Mn^{2+}, Al)_2Li_3Si_{12}O_{30}$

Composition	silicate
Crystal system	hexagonal
Hardness	6–6.5
Specific gravity	2.74
Streak	white
Lustre	vitreous

Description Most commonly found as solid lumps and coatings, this mineral is a light to very dark purple colour. It has several common names, such as wesselite, after the Wessels mine where it originates, and royalazel, referring both to its royal purple colour and to Hotazel, a town near to the Wessels mine.

Uses Used for stone carving and lapidary purposes. Hundreds of tons have been produced over the past 30 years, varying in quality from an expensive and translucent deep purple variety to pale purple opaque material. Rare transparent crystals are in demand for use as faceted gemstones.

Occurrence South Africa NC: The Wessels mine, Kalahari manganese field, produces masses of sugilite of up to a kilogram or more, often interlayered with fine-grained aegirine, pectolite and manganese oxides. More rarely, the mine produces world-class sugilite crystals of up to 15 mm.

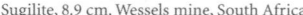

Sugilite, 8.9 cm. Wessels mine, South Africa.

Rough sugilite with tumbled beads of green chrysoprase and sugilite, 11 cm. Kalahari manganese field, South Africa.

Talc

$Mg_3Si_4O_{10}(OH)_2$

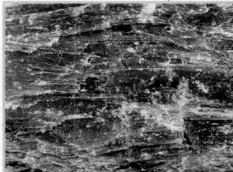

Composition	silicate
Crystal system	monoclinic
Hardness 1	
Specific gravity	2.58–2.83
Streak	white
Lustre	dull, pearly

Description One of the softest minerals known, talc can be scratched with a finger nail. It is found in metamorphosed siliceous dolomites and hydrothermally altered ultramafic rocks, i.e. rocks through which hot geothermal fluids have passed, breaking down and chemically altering the minerals. It is commonly referred to as a 'soapstone' because of its greasy, soapy texture.

Uses It is used in the cosmetic and pharmaceutical industries, but also has industrial applications as a paint additive, and in ferrocastings and ceramics. It is very easily fashioned into stone carvings and other lapidary items.

Occurrence **South Africa** LIM: Occurs in the Pietersburg and Murchison greenstone belts. MP: Found near Nelspruit. Several talc mines have operated in the Barberton district. GAU: Occurs in the Honeydew, Muldersdrif and West Rand regions. KZN: Talc is found in metamorphic marbles at Marble Delta and in the Nqutu district in the Tugela River Valley. **Namibia** Occurs in the Windhoek and Gobabis districts. It is also found in dolomite east and southeast of Windhoek. **Botswana** Talc schists outcrop in the eastern part of Botswana. **Zimbabwe** Large volumes have been mined from the Athi mine, Bubi district, and the Gray mine in the Nyanga district. Other deposits are exploited in the Kwekwe, Makoni and Wedza districts. **Swaziland** Economic deposits of talc occur 3 km west of Sicunusa, in chlorite and chlorite-carbonate schist.

Talc, 18 cm. Barberton district, South Africa.

Tantalite-(Fe)

$Fe^{2+}Ta_2O_6$

Composition	oxide
Crystal system	orthorhombic
Hardness	6–6.5
Specific gravity	8.2
Streak	tantalite-(Fe) brown, black
Lustre	tantalite-(Fe) weakly metallic

■ **Description** 'Tantalite' was a general name for the tantalite-(Fe) and tantalite-(Mn) series, named after the presence of tantalum in the minerals. Tantalite-(Fe) is the more common variety. It can form metallic black euhedral crystals, although shapeless lumps are usual.

■ **Uses** Tantalum has an extremely high melting point of almost 3 000 °C. It is used in the manufacture of certain electronic and chemical products such as oscillators, amplifiers and alloys.

■ **Occurrence** **South Africa** LIM/MP: Occurs as large crystals in some pegmatites, e.g. at Pala Kop west of Giyani. **Namibia** In the Karibib-Usakos region tantalite-(Fe) is found in pegmatites bearing rare-earth elements and tin. Tantalite Valley in southern Namibia is named for the extensive deposits of this mineral. The pegmatite region between Karibib and Brandberg West also contains tantalite-(Fe) mineralization. At the Rubikon mine, crystals up to 3 cm were found, and at the Helikon mine, crystals up to 16 cm in diameter occur. **Zimbabwe** Over 100 pegmatites in Zimbabwe have been exploited for tantalite-(Fe). It is commonly associated with beryl, cassiterite and lepidolite and most deposits are in the Hurungwe and Mutoko districts. **Swaziland** Alluvial gravels in the Forbes Reef area contain tantalite-(Fe), as do the tin-bearing gravels near Mbabane and those at the Star mine near Sinceni.

Tantalite-(Fe), 9.5 cm. Grietjie mine, South Africa.

Titanite
$CaTiSiO_5$

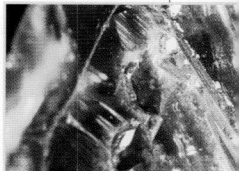

Composition	oxide
Crystal system	monoclinic
Hardness	5–5.5
Specific gravity	3.4–3.5
Streak	white
Lustre	adamantine

Description Previously named sphene. Titanite forms wedge-shaped or prismatic crystals that are characteristically yellow, orange or brown, but that can also be grey or black. It occurs in granites or metamorphosed calc-silicate rocks.

Uses It can be a titanium ore when it occurs in heavy-mineral beach sand. (See zircon and rutile.) Occasionally it is used as a novelty stone in jewellery.

Occurrence South Africa NC: Twinned yellow ('fishtail') crystals up to 3 cm have been found in the Richtersveld. MP: It occurs in granites in the Pilgrim's Rest district, near Mt Anderson. NW: Found in the Pilanesberg rocks and in the Goudini carbonatite. **Namibia** Large titanite crystals up to 9 cm come from the farm Tantus in the Gamsberg region. Also found at the Krantzberg mine and in carbonatites in the Kaokoveld Complex and the Ais dome skarn north of Uis. **Zimbabwe** Found as an accessory mineral in the Shawa and Dorowa carbonatites. Gem-quality yellow titanite came from the Ju Jube mine in the Hurungwe district, 4 km east of the Ball mine, Mutoko, and in calc-silicate rocks in the Makuti area.

'Fishtail' twinned gem-quality titanite crystal, 2 cm. Ju Jube mine, South Africa.

Titanite crystal, 3.4 cm, in granite. Pilgrim's Rest district, South Africa.

Topaz
$Al_2SiO_4(F,OH)_2$

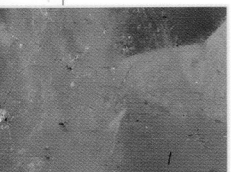

Composition	silicate
Crystal system	orthorhombic
Hardness	8
Specific gravity	3.49–3.57
Streak	white
Lustre	vitreous

Description Topaz is found in granitic pegmatites often associated with the other common pegmatite minerals such as quartz, microcline and schorl tourmaline. The crystals are hard and, if transparent, facetable. Their colour varies from glassy and colourless to pale blue, blue, amber, yellow or honey-coloured.

Uses Topaz has been used as a gemstone for centuries. 'Imperial topaz' is amber coloured and is so-called because it was used in the jewellery of the Russian royal family. High-quality dark blue topaz crystals have been mined in northwest Zimbabwe and large quantitites of gem-quality colourless topaz have been mined for many decades in Namibia.

Large topaz crystal, 10.9 cm. St Ann's mine, Zimbabwe.

Occurrence **South Africa** NC: Topaz is recorded from a few pegmatites in the province. GAU: A small cassiterite-bearing greisen outcrop occurs about 90 km northeast of Pretoria. **Namibia** Klein and Gross Spitzkoppe have produced 'silver topaz' – colourless transparent crystals. Pale blue and colourless topaz crystals come from the Erongo mountains, Karibib and Usakos pegmatites. The Brandberg Mountain has yielded topaz crystals weighing several kilograms. **Zimbabwe** Beautiful blue topaz crystals over 10 cm are mined at the St Ann's mine, Mwami area, in the Hurungwe district – a world-famous topaz locality. Also found in the ancient alluvial Somabula deposits, Gweru district. **Swaziland** Topaz is found in pegmatites in the foothills of the Sinceni mountains.

'Silver topaz', 3.6 cm. Klein Spitzkoppe, Namibia.

'Imperial topaz', 4.8 cm. Zambia.

Tourmaline group – see elbaite and schorl.

Vanadanite

$Pb_5(VO_4)_3Cl$

Composition	vanadate
Crystal system	hexagonal
Hardness	3
Specific gravity	6.88
Streak	white to pale yellow
Lustre	resinous, subadamantine

■ **Description** A member of the apatite group of minerals, vanadanite forms perfect hexagonal crystals that are elongate and barrel-shaped or flat and tabular. They are generally vibrant red in colour, but yellow-brown, amber and brown-red shades also occur.

■ **Uses** Vanadium is a refractory metal. It is used in the steel industry and as a blue and yellow pigment in ceramics and glass, as well as in superconductors and fuel cells. It also serves as a drying agent in ink, paint and varnish.

■ **Occurrence** **South Africa** GAU: Occurs at the old Argent mine, Delmas district. NW: Crystals up to 1 cm have been found on the farm Kafferskraal 306 JP, Ottoshoop district. It is also found at the Kindergoed deposit, Marico district. **Namibia** The largest vanadanite crystals in the world came from the Abenab mine, Otavi mountainland (many giant cigar-sized crystals). Crystals were also found at the Namib Lead mine. **Zimbabwe** Vanadanite crystals occur at the OBE mine, near Bulawayo.

Vanadanite crystals on pyromorphite, 2.3 cm. North West, South Africa.

Vanadanite, 2.1 cm. Namib Lead mine, Namibia.

Wulfenite

PbMoO$_4$

Composition	molybdate
Crystal system	tetragonal
Hardness	2.75–3
Specific gravity	6.5–7
Streak	white
Lustre	adamantine, resinous

Description This is a relatively rare mineral, but spectacular wulfenite crystals – popular with collectors worldwide – come from Namibia. It forms in the secondary oxidation zones of certain lead deposits as distinctive, extremely flat, tabular crystals with bevelled edges. It is usually bright yellow or orange.

Occurrence **South Africa** GAU: Rare, but small crystals have been reported from the Leeuwenkloof and Argent mines. **Namibia** Some of the finest wulfenite crystals in the world have been found at the Tsumeb mine, in the Otavi mountainland. These range in colour and may be grey to cream, canary yellow, red or, rarely, blue. Some are 2.5 cm thick and 5 cm on edge. Light caramel-coloured wulfenite crystals were found at the Khuiseb Springs deposit, to the west of Tsumeb. **Zimbabwe** Wulfenite is found at the Osborne's wulfenite deposit, Gutu, and in the Selukwe region.

Tabular wulfenite crystal, 2.3 cm. Tsumeb mine, Namibia.

Wulfenite and quartz, 4 cm. Tsumeb mine, Namibia.

Zircon

ZrSiO$_4$

Composition	silicate
Crystal system	tetragonal
Hardness	7.5
Specific gravity	4.6–4.7
Streak	white
Lustre	vitreous, adamantine

Description Zircon crystals are prismatic and vary in colour, including caramel, tan, brown, and red-brown to black shades. They are usually opaque, but sometimes translucent to transparent. Most commonly found in pegmatites and kimberlites and in heavy-mineral beach sands (i.e. coastal environments where waves rework the sand and concentrate the heavy minerals in layers).

Uses Zircon sands are used as a refractory product, and in mouldings and some superconductors. Zirconium is extracted from zircon and is used in control rods in nuclear power stations. It is also used in glass, fuel cells, deodorants and ceramics. Cubic zirconia, which is a manufactured gemstone, is unrelated to zircon.

Occurrence **South Africa NC/LIM/NW:** Relatively common in some kimberlite pipes. **KZN:** Large brown crystals occur along the Umhlatuzi River on Bull's Run Estate 12987. Zircon is a major component of the heavy-mineral beach sands at Richards Bay. These sands also contain ilmenite, rutile, monazite and garnet. **WC:** In the Vanrhynsdorp district, zircon and monazite deposits occur at Steenkampskraal, Uitklip and Roodewal. **Namibia** Euhedral crystals occur in

Zircon, 2.8 cm. Bull's Run Estate, South Africa.

pegmatites, e.g. in the Karibib district. Zircon is an accessory mineral in kimberlites in the Gibeon district. **Zimbabwe** It is a common accessory

Zircon, 4 cm. Zomba Mountain, Malawi.

mineral in granites, with large crystals occurring in pegmatites. **Swaziland** Zircon is found in granites and some pegmatites in Swaziland. It is also found in alluvial gravels east of the Lebombo Mountain. **Mozambique** Zircon is found in beach sands as well as in dune deposits on the southern Mozambiquan coastline.

ROCKS

All rocks are classified into three groups – igneous, sedimentary and metamorphic. The table below lists the main characteristics of each group:

Classification of rocks		
Igneous	**Sedimentary**	**Metamorphic**
Crystallization from a molten magma or molten lava.	Deposited by water, wind, ice or gravity. Organic deposits. Evaporites (minerals formed by evaporation). Chemical precipitates.	Re-crystallization from pre-existing rocks, and melts.
Examples		
basalt	chert	gneiss
dolerite	coal	hornfels
gabbro	halite	marble
granite	limestone	schist
rhyolite	mudrock	slate
	sandstone	

Because the Earth is a dynamic, evolving planet, rocks that formed millions of years ago may not exist today or may have been transformed by plate tectonic processes.

The impressive granite mountain of Gross Spitzkoppe, Namibia.

The diagram below illustrates the cyclical evolution whereby rocks are created, destroyed or transformed:

The rock cycle

Magma reaches the Earth's surface, erupts as lava, cools, crystallizes and solidifies to form igneous rock. Via erosion and weathering, sediments are transported to new locations and deposited in layers. With increasing deposition and plate tectonic subduction, these layers of sediments become buried and lithified into hard rocks. At this point tectonic activity may cause the layers to be uplifted and the cycle to begin again.

However, if sedimentary rock becomes more deeply buried, the immense pressure and heat may cause it to alter composition and become metamorphic rock or, at great depths and pressure, even to melt.

This cycle takes place over millions, or even hundreds of millions, of years. Once sediment is deposited (top right), it takes a long time for it to be buried and slowly transformed into hard sedimentary rock. It takes even longer for sedimentary rock to become so deeply buried that it undergoes metamorphism. Greatest of all is the time taken for rocks formed deep within the Earth to make their way to the surface by means of tectonic uplift, thereby beginning the cycle again. As a result of this immensely slow process, continents and oceans are created and destroyed, and this has happened on a continuous basis throughout the Earth's 4.5 billion-year history.

Igneous rocks can also form when primary, molten magma in the Earth's mantle rises up into the crust and crystallizes there.

Igneous rocks

Igneous rocks form when either molten magma (ascending from deep in the Earth) or lava (flowing on the surface of the Earth) cools and crystallizes to form minerals. The key difference between these two broad categories of igneous rocks is that, in the former, the mineral crystals are large enough to be visible to the naked eye, while the latter comprises very small, fine-grained minerals. This difference is a function of the cooling rate of the molten material – the quicker it cools, the smaller the crystals.

The liquid 'melt' source material for igneous rocks originates in the Earth's mantle. It is made up of a mixture of elements and so different types of minerals can crystallize from it, depending on the chemical and temperature conditions. The silica content is important: a melt that is silica-poor tends to be hotter; a melt that is rich in silica tends to be cooler.

High-temperature (900 °C), silica-poor melts generally contain the elements iron, magnesium and calcium. These crystallize to form minerals such as olivine, pyroxene and plagioclase feldspar respectively. As the magma gradually cools, different minerals follow in a fixed crystallization sequence (summarized in Bowen's Reaction Series). Lower temperature melts contain more potassium and silica. The minerals that commonly form in these chemical and temperature conditions include quartz and alkali feldspars (like orthoclase and microcline). If the melt ascends slowly from the mantle to the crust, cooling very gradually, this can produce a wider range of minerals and allow larger crystals to form.

The granite mountains of Klein Spitzkoppe, Namibia.

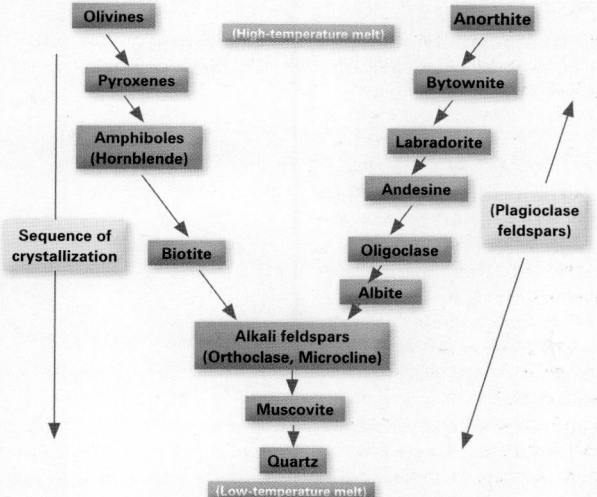

Bowen's Reaction Series

Bowen's Reaction Series illustrates the order in which rock-forming minerals crystallize from a melt.

Olivines

Anorthite

(High-temperature melt)

Pyroxenes

Bytownite

Amphiboles (Hornblende)

Labradorite

Andesine

(Plagioclase feldspars)

Sequence of crystallization

Biotite

Oligoclase

Albite

Alkali feldspars (Orthoclase, Microcline)

Muscovite

Quartz

(Low-temperature melt)

An igneous rock such as a peridotite, composed almost entirely of olivine, is a high-temperature rock. A granite, made up of quartz, alkali feldspar and muscovite mica, is a lower-temperature rock.

Classification of common igneous rocks

Grain size	Silica-rich	Intermediate	Mafic	Ultramafic (Silica-poor)
Coarse (> 5mm)	Granite	Granodiorite / Diorite	Gabbro	Peridotite
Fine (< 1mm)	Rhyolite	Dacite / Andesite	Basalt	Komatiite

Igneous rocks are classified according to the size of their constituent minerals (grain size) as well as their overall chemical composition, usually related to silica content. Grain size reflects the setting in which the rocks formed, i.e. either intrusive, meaning that magma rose up and penetrated existing rock layers below the Earth's surface, solidifying there, or extrusive, meaning that the magma erupted from a volcano and cooled on the land's surface. Examples of different categories of coarse-grained rocks and their fine-grained equivalents are given above.

Anorthosite

Colour Anorthosite is usually a pale to white rock because of the abundance of plagioclase feldspar in its composition. It often has a speckled appearance, caused by scattered dark minerals such as pyroxene.

Composition Anorthosite is a coarse-grained igneous rock made up of more than 90% plagioclase feldspar, labradorite, bytownite or anorthite.

Uses Labradorite feldspar has an attractive blue iridescence and is quarried for building purposes.

Occurrence South Africa Thick layers of anorthosite occur in the Bushveld Complex. Anorthosite is also found in Namaqualand and close to Aggeneys, where it contains ruby corundum. **Namibia** The Kunene Complex contains large volumes of anorthosite. The mineral is closely

Anorthosite, 9.5 cm. Bushveld Complex, South Africa.

associated with sodalite deposits at Swartbooisdrif in Kaokoland, and is also found in the Sinclair Sequence. **Botswana** Some anorthosite occurs in eastern Botswana. **Zimbabwe** In the Great Dyke it is associated with economic chromite deposits.

Basalt

Colour Dark grey to black. May be vesicular (i.e. has hollows) or contain light-coloured amygdales filled by secondary minerals such as quartz, calcite, chalcedony, agates and other minerals, e.g. zeolites. These mineral-filled amygdales can be circular or form elongate, pipe-like features.

Composition Basalt is the extrusive volcanic equivalent of gabbro or dolerite and therefore has the same mineral composition as these mafic rocks.

Uses Basalt is sometimes used as dimension stone.

Occurrence South Africa/ Lesotho The most widespread basalts in southern Africa are the Jurassic-Cretaceous flood basalts

Amygdaloidal and vesicular basalt, 6.5 cm. Drakensburg, South Africa.

that form the bulk of the upper Drakensberg. Basalts also outcrop in northern KwaZulu-Natal, Springbok Flats, the Waterberg, the extreme eastern border between South Africa and Mozambique, and northern Limpopo. **Namibia** Similar types of basalt occur in the Etendeka lavas in northwest Namibia and basalts are widespread in the Goboboseb Mountains at Tafelkop, west of the Brandberg. **Zimbabwe** 'Drakensberg-type' basalts are also found in western and northwestern Zimbabwe. An amygdaloidal basalt from the Save River Valley forms part of the Umkondo volcanics and is used as dimension stone. **Swaziland** Basalts outcrop in the eastern parts of the country.

Carbonatite

Colour Carbonatites are dark to mottled grey-black, but may be off-white if abundant calcite is present. They are often coarse-grained.

Composition These rocks have a high calcium carbonate (calcite) content and frequently contain magnetite, pyroxene, diopside and several accessory minerals such as zircon and apatite-(CaF).

Uses Carbonatites can host copper deposits, and some are also exploited for phosphate and zirconium. Rare-earth element minerals (a group of 30 metallic elements that fall in Group 3 on the periodic table) may be present in some carbonatites.

Occurrence The modes of occurrence and mineralogy of carbonatite may resemble those of kimberlite. Both form pipe-like features that intrude into the surrounding rock, and many carbonatites have a large ultramafic component. **South Africa** The Phalaborwa carbonatite is best known. Other carbonatitic localities include the Salpeterkop pipe in the southern Karoo and Schiel, Stukpan, Spitskop, Tweerivier, Kruidfontein, Nooitgedacht, Goudini, and Glenover. **Namibia** Dicker Willem (between Aus and Lüderitz), Kalkveld, Okorusu, Otjisazu, Ondurakorume, Marinkas Quellen and Gross Brukkaros are well-known carbonatites – most of these are concentrated in southern and north-central Namibia. **Botswana** Carbonatites occur in southwest Zimbabwe where three large carbonatite bodies – Dorowa, Shawa and Chishanya – are found.

Carbonatite composed of calcite, diopside, magnetite and apatite-(CaF), 14 cm. Palabora mine, South Africa.

Dolerite/Diabase

Colour Dolerite is dark grey to black and is commonly speckled.

Composition It is the intrusive equivalent of a gabbro, i.e. the chemical and mineral composition is the same. It consists of calcic plagioclase and pyroxene, with minor quantities of other mafic minerals such as olivine or ferrohornblende. Dolerite is finer grained than gabbro because it cooled faster while intruding into the surrounding rocks. Diabase is a synonym for dolerite. In southern Africa, diabase usually refers to dolerites that are older than the Jurassic-age Karoo dolerites.

Uses Dolerite is used as dimension stone. It is also widely quarried and used for road construction and as aggregate in the manufacture of concrete.

Occurrence Dolerites occur in sills (sheet-like bodies of rock that intrude horizontally) and dykes (vertical or inclined features that cut across the host rock). These intrusions vary tremendously in size. While dykes are often only a few metres thick, some sills can outcrop for kilometres and be hundreds of metres thick. Most dolerite is fine-grained, but some thick sills have coarse inner zones with outer chill margins made up of very small crystals, i.e. the thicker inner region cooled too slowly for fine grains to form, while those areas directly in contact with the surrounding rocks cooled much faster allowing a finer grain to develop. Many dolerites make good aquifers. Dolerites are relatively common in **South Africa**, **Namibia**, **Botswana** and **Zimbabwe**. The most easily recognizable dolerites are those found in the Karoo region in South Africa, where sills formed very characteristic flat-topped koppies.

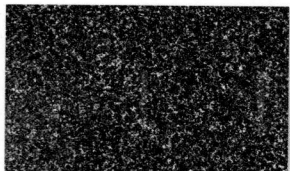

Dolerite, composed of dark pyroxene and lighter coloured plagioclase feldspar, 22 cm. Piet Retief district, South Africa.

Weathered and eroded dolerite sill exposed by the Vaal River. Barkly West district, South Africa.

Because dolerite is harder than the surrounding sedimentary rock, it resists weathering and thus forms flat caps to hills. Slopes strewn with weathered dolerite boulders are common features in the southern African landscape. The boulders tend to be sub-rounded and brown to red-brown, caused by the oxidation of the ferromagnesium minerals.

■ Gabbro, Norite

■ **Colour** Both gabbro and norite are dark, usually black, with some white flecks of plagioclase feldspar. It is difficult to distinguish between hand specimens of these two rocks.

■ **Composition** A gabbro is a plutonic, mafic igneous rock that is composed of plagioclase feldspar (labradorite to anorthite), and clinopyroxene, usually enstatite or augite, and some olivine (see Bowen's Reaction series). Accessory minerals may include ferrohornblende and biotite, as well as magnetite and ilmenite. Norite is similar to gabbro but contains a different orthopyroxene, hypersthene (enstatite).

■ **Uses** Gabbro and norite are quarried extensively, for use locally and for the export market. They are used for tiles, tabletops, tombstones and to clad the exterior walls of buildings.

■ **Occurrence** **South Africa** Common in Namaqualand and the Bushveld Complex (a huge intrusive complex about 530 km long and 150 km wide). Gabbro and norite occur layered between other ultramafic rocks, namely dunite and pyroxenites. Large thicknesses of the dunite have been serpentinized. **Namibia** Gabbro and norite occur in the Sinclair Sequence and other areas. **Botswana** Both these rocks are found at Molopo Farms in southern Botswana. **Zimbabwe** They occur in the Great Dyke, associated with chromite. **Swaziland** Gabbro is found together with pyroxenites in the Usushwana Complex.

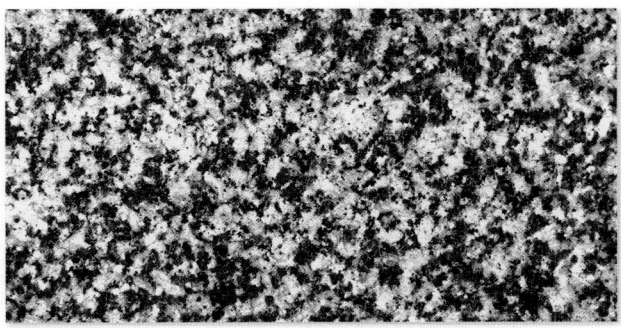

Gabbro, 10 cm. Bushveld Complex, South Africa.

Granite, Granite pegmatite

Colour   Granite varies in colour from pink-red, if orthoclase is abundant, to white, if quartz and microcline dominate the mineral assemblage in the rock. *Granite pegmatite* is light in colour, ranging from white to cream, but can contain a range of exotic, coloured minerals such as tourmaline, cassiterite, or the rare-earth element minerals.

Composition   Granite is coarse-grained plutonic igneous rock: its crystals are usually 2 mm or larger. Two minerals predominate: quartz and potassium-bearing feldspar (particularly orthoclase and/or microcline). Some sodium-bearing plagioclase feldspar, albite, muscovite, biotite and black ferrohornblende may be present. Granite weathers in a distinctive pattern. Layers of rock weather and crumble by exfoliation (peeling like the layers of an onion), leaving hills and koppies made up of enormous rounded to sub-rounded boulders. *Granite pegmatite* has the same composition as granite, but a coarser grain size. Some granitic pegmatites occur in veins or isolated pockets that often outcrop as white ridges on the Earth's surface. They may contain crystals metres long and weighing several tons.

Uses   Granite is used extensively as dimension stone. Gemstones such as tourmaline and topaz are concentrated in granitic pegmatites, as are other economic minerals such as ferberite, cassiterite and scheelite. Pegmatites are therefore mineralogically very interesting rocks and they are often the source of esoteric minerals and gemstones.

Occurrence   In southern Africa, some of the oldest rocks, over 3 000 million years old, are granites and granitic pegmatites. Granitic pegmatites that host valuable minerals and gemstones are found in central Namibia, the Northern Cape, South Africa and northwest Zimbabwe. **South Africa** Granite occurs in the Northern Cape (Namaqualand), northern Limpopo and eastern Mpumalanga. In the Western Cape granite is relatively common (e.g. Paarl Mountain). **Namibia** The Spitzkoppe and Brandberg mountains are famous granitic landmarks. Central Namibia has many granite deposits and some have been quarried. **Botswana** Granite outcrops in the Gaborone district.

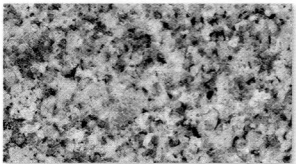

Granite composed of quartz, microcline, biotite and albite, 8 cm. Klein Spitzkoppe, Namibia.

Coarse-grained granitic pegmatite with large quartz and microcline crystals, 45 cm. Klein Spitzkoppe, Namibia.

Quarrying granite in the Bushveld Complex, South Africa.

Zimbabwe Most of the eastern half of the country consists of granite. The Matopos are also composed of granite. **Swaziland** Almost two-thirds of the country is underlain by granite and granite-gneiss.

Kimberlite

Colour Kimberlite varies in colour from grey to brown to blue. To identify kimberlite in the field, one looks for a blue-coloured rock, which has given rise to kimberlite's common name 'blue ground'. It has a typical mottled appearance with a mixture of grain sizes. Kimberlite is sometimes also referred to as a brecciated peridotite.

Composition The mineralogy of kimberlites can be complex but the rock consists mainly of apatite-(CaF), calcite, diopside, ilmenite, monticellite, olivine, phlogopite, perovskite and serpentine. Other minerals sometimes present are zircon, garnet and spinel. Kimberlite often contains xenoliths (foreign rock fragments) from the upper mantle composed of peridotite and eclogite, and xenocrysts (foreign minerals) such as chrome diopside and chromian spinel. Diamonds are rare xenocrysts in kimberlite. Kimberlites are broadly subdivided into two groups, so-called Group-I, olivine-rich monticellite-serpentine-calcite kimberlites, and Group-II, which are mica-rich.

Uses Some kimberlites are mined for diamonds.

Kimberlite, 13.5 cm. Du Toitspan, South Africa.

Occurrence **South Africa** Kimberlite occurs in small volcanic pipes, dykes and sills. Over 800 kimberlites are known, but only about 50 contain diamonds. Kimberlites range in age from 80–160 million years old (the Cretaceous to Late Jurassic Period) to 1 640–1 720 million years (the Proterozoic Eon). The kimberlites at the Premier mine are 1 200 million years old – chronologically the oldest diamond-bearing pipe in the world. Many world-famous gem-quality diamonds have been discovered in South Africa, but the kimberlites also contain other minerals that are of interest to collectors. **Namibia**, **Botswana**, **Zimbabwe**, **Swaziland**, and **Lesotho** all have kimberlites, but Namibian kimberlites do not contain diamonds.

Peridotite

Colour Characteristically black or dark green-black, peridotite can also have scattered light-coloured patches.

Composition A coarse-grained ultramafic igneous rock, peridotite consists mainly of olivine and other dark ferromagnesium minerals such as pyroxene. There are several varieties of peridotite, including dunite (which is composed mainly of olivine) and pyroxenite (which consists mainly of pyroxenes).

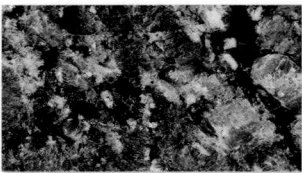

Peridotitic pyroxenite composed of dark green pyroxene and anorthite, 15 cm. Bushveld Complex, South Africa.

Komatiites are a special type of ultramafic peridotite containing very high percentages of MgO and are unique to the ancient greenstone belts. Kimberlites, which contain diamonds, are also a variety of peridotitic rock.

Occurrence **South Africa** Pyroxenites are common in the basal rocks of the Bushveld Complex. Some Bushveld Complex rocks have ultramafic pegmatites composed of large pyroxene crystals, up to 15 cm long. **Namibia** Peridotite is known from the Kunene Complex. **Botswana** Pyroxenites and peridotites occur in Molopo Farms, in the east. **Zimbabwe** Pyroxenites occur in the Great Dyke. **Swaziland** The Usushwana Complex located southwest of Mbabane consists of pyroxenite and gabbro.

Porphyry

Colour The colour of porphyritic rocks varies. The typical spotted appearance is a result of their unusual composition.

Composition 'Porphyry' is the general name used for igneous rocks that have a bimodal mineral size distribution, i.e. a fine-grained matrix that contains larger crystals of other minerals, referred to as phenocrysts.

Quartz and alkali feldspars are common phenocrysts. Porphyrys have a dual mode of origin and this accounts for the bimodal crystal sizes in the rock. Magma first cools slowly at depth, and large phenocryst crystals form. Then, while the magma is still partially melted, with the larger phenocrysts floating in the mass, it either intrudes or erupts as lava. Both of these processes rapidly cool the remaining magma, which forms a fine-grained groundmass enclosing the larger phenocrysts.

Porphyry composed of fine matrix containing large feldspar phenocrysts, 19 cm. Ventersdorp Supergroup, South Africa.

■ **Uses** Porphyritic rocks can be used ornamentally in building.

■ **Occurrence** **South Africa** Porphyry is relatively common in the volcanic sequence of the Ventersdorp Supergroup. It also occurs in association with alkaline intrusives scattered throughout southern Africa. **Namibia** Porphyrys are found in the Sinclair Sequence.

■ Rhyolite

■ **Colour** Rhyolite is a fine-grained, silica-rich, cream to off-white, igneous rock.

■ **Composition** Rhyolite is the extrusive equivalent of granite. It therefore has a similar chemical composition and mineral assemblage to granite, namely quartz, alkali feldspar, muscovite and minor biotite. Rhyolite lavas often display a distinct layered or banded texture.

■ **Uses** It is occasionally used as an ornamental building stone.

■ **Occurrence** **South Africa** The Ventersdorp Supergroup contains rhyolite, as does the Bushveld Complex Rooiberg Group. Rhyolite also occurs in the Lebombo mountains of northern KwaZulu-Natal and extends into southern **Swaziland**.

Rhyolite, displaying typical layering, 7 cm. Swaziland.

■ Syenite

■ **Colour** Syenites are often red because of the presence of orthoclase feldspar. They resemble granite, but contain no quartz.

■ **Composition** Syenite is a coarse-grained intermediate igneous rock that contains alkali feldspars and/or feldspathoids, such as nepheline. Various syenites occur – usually named after the dominant mineral present. Some syenites may contain quartz, others not. Zircon, titanite and apatite-(CaF) are common accessory minerals. Syenites are common in some carbonatite complexes.

■ **Occurrence** **South Africa** Syenites are relatively common in some carbonatites such as the Phalaborwa Complex and in the Pilanesberg region in the North West province. They also occur in other carbonatites in the country. **Botswana** Syenites occur west of Jwaneng and in the Kweneng district. **Zimbabwe** Syenite is found in the Dorowa and Shawa pipes.

Syenite, 11.2 cm. Pilanesberg, South Africa.

Sedimentary rocks

Sedimentary rocks are subdivided into two main groups:

- **clastic sedimentary (or allochthonous) rocks**, and
- **non-clastic sedimentary (or autochthonous) rocks**.

These two groups contain several different categories, defined by their composition, their mode of origin, or both.

In the *clastic/allochthonous* group, sediment is composed of clay, silt, mud, sand and gravel, i.e. the weathered material from pre-existing rocks. It can also be formed by the deposition of volcanic debris such as volcanic ash. The term sediment refers to loose, unconsolidated material, while the term sedimentary rock refers to the lithified, hard rock. ('Sediment' should not be used when describing the rocks themselves.)

In the *non-clastic/autochthonous* group, the chemical and organic sediment includes material formed by evaporation. Examples are halite (rock salt), limestone formed from the skeletal remains of carbonate organisms, coal formed from plant material, and other chemical sediments such as chert and iron formations.

This classification of sedimentary rocks can be somewhat simplified by answering the question: was the sediment transported and then deposited, or was the sediment formed/crystallized *in situ*?

Clastic sedimentary rocks (allochthonous)	Non-clastic sedimentary rocks (autochthonous)
Mudrock	Limestone
Siltstone	Coal
Sandstone	Dolomite
Conglomerate	Chert
Tuff	Halite
Agglomerate	Banded-iron formations

Weathered sandstone exposed on a beach, Eastern Cape, South Africa.

Outcrops of the Clarens sandstone in the Eastern Cape, South Africa. This geological formation is well known for its caves.

Clastic sediment and clastic sedimentary rocks are classified according to the size of the mineral grains that make them up. The very smallest clastic particles are clay (less than 0.0039 mm diameter), followed by silt-sized particles (0.038–0.063 mm) and sand particles (0.063–2 mm). Gravel comprises any particles larger than 2 mm diameter and may include boulders several metres in size.

Grain size classification of clastic sedimentary rocks

Grain size in millimetres	Sediment		Sedimentary rock	
	Components			
>256	Boulders	Gravel	Conglomerate	
16–256	Cobbles			
4–16	Pebbles			
2–4	Granules		Granulestone	
0.063–2	Sand		Sandstone	
0.0039–0.063	Silt	(Mud)	Siltstone	Mudrock
<0.0039	Clay		Claystone	

Non-clastic sedimentary rocks are classified according to their composition.

Classification of non-clastic sedimentary rocks

Composition	Examples
Chemical precipitates	Chert Halite
Organic deposits	Coal Limestone
Residual deposits (formed by *in situ* weathering)	Laterite (iron-rich residual soil) Bauxite (aluminium-rich residual soil)

Chert

Colour Chert is typically grey to grey-white, or dark grey to black, but can be coloured by impurities, e.g. red jasper is iron-rich chert. Flint is brown chert.

Composition Chert is extremely fine-grained (crypto-crystalline) silica. It is hard and has a conchoidal fracture pattern. Chert can form either organically or inorganically as a primary deposit, from the deposition of silica-rich fluids or from siliceous oozes. It may also form by secondary replacement when percolating silica-rich fluids infiltrate existing strata. Banded-iron formations are rocks that are composed of layers of alternating iron-rich chert (jasper) and iron-poor chert.

Uses It is sometimes used as an ornamental stone or for jewellery. Chert was often used by early man to make primitive tools such as hand axes and arrowheads.

Occurrence Chert is widely distributed in dolomite in **Namibia**, **Zimbabwe** and **South Africa**, where it forms replacement lenses or layers in the dolomites. In South Africa, large deposits of jasper occur in the Northern Cape, associated with the iron and manganese deposits there. In Zimbabwe, banded iron-formations – called jaspilite – are found in the Midlands. Jaspilite is red-brown, banded and/or brecciated, and therefore makes very attractive ornamental stones and lapidary material. Chert occurs in limestones and dolomites in the Lomagundi, Piriwiri and Bulawayon group rocks.

Fresh grey chert layer within weathered brown chert, 3 cm. Barberton district, South Africa.

Coal

Colour Black. Usually banded with alternating bright and dull layers.

Composition Coal originates from peat – organic plant material that accumulates in swamps and marshes. If layers of peat are protected from oxidation and get covered over by sediment and buried, the peat gradually transforms into hard coal. This is because burial leads to compaction and dewatering, and increases in geothermal heat and pressure, which changes the chemical composition of the peat. Coal is classified according to its rank, type and grade.

Rank reflects the effect of increasing pressure being applied to the decaying plant matter. The highest ranked coal, anthracite, has the lowest water and the highest carbon content (see illustration).

Type reflects the kinds of organic material – known as macerals – that comprise the coal. The main macerals are vitrinite, exenite and inertinite, each being a different type of plant material.

Grade is based on the presence of inorganic impurities in the coal. Common impurities are clay minerals and quartz that have washed into peat swamps. Pyrite is a highly undesirable mineral in coal as it produces sulphur dioxide when burnt, contributing to pollution and acid rain. Coking coal is a special grade used in the manufacture of steel. It swells when heated, which increases its surface area. Thus, when air passes through the coal, great heat is generated.

Bituminous coal, 10.8 cm. Witbank coalfield, South Africa.

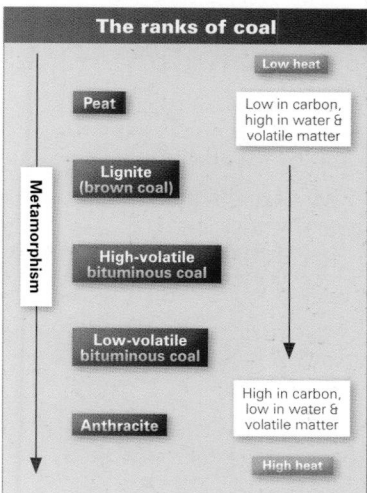

The ranks of coal

Metamorphism

Peat

Lignite (brown coal)

High-volatile bituminous coal

Low-volatile bituminous coal

Anthracite

Low heat — Low in carbon, high in water & volatile matter

High in carbon, low in water & volatile matter — High heat

■ **Uses** Coal is used primarily for electricity generation, for synthetic fuel (synfuel) production, metallurgical processes and domestic heating.

■ **Occurrence** All of the southern African countries, except Lesotho, have coal deposits of varying size and quality. The main coal producers are South Africa, followed by Zimbabwe, Botswana, Mozambique and Swaziland. Although Namibia has coal, no coal mining takes place at present. In **South Africa** the major coalfields are centred in an arc extending from Vereeniging in the west, through Grootvlei, Delmas, eMalahleni, Middelburg, Emakhazeni and into northern KwaZulu-Natal. Much of the coal is fairly close to the surface and can be exploited via opencast mining methods. This coal is mainly bituminous, although anthracite is mined in

KwaZulu-Natal. Large deposits are also found in the Springbok Flats, Waterberg and Soutpansberg areas. Coal was first discovered and mined in the Eastern Cape, in the Molteno district. In **Namibia**, extensive coal deposits are found in the Aranos coalfield, although these are fairly deeply buried. Some coal is found in the Ovamboland, Toscanini and Waterberg areas. In **Botswana** several coalfields are known, one of which, the Morupule deposit, is being exploited for power station feedstock. Coal also occurs in the Kweneng, Mamabula, Tuli and northeast sections of Botswana. In **Zimbabwe**, coal has been mined for decades in the Hwange district at Wankie in the western part of the country. Other coal deposits occur in the Sabie Valley, Beitbridge and Binga districts. The Tete coalfield in **Mozambique** has been periodically exploited for bituminous coal as well as anthracite. **Swaziland** has one operating coal mine, Maloma, where anthracite is exploited.

■ Conglomerate

■ **Colour** Variable but always displays a variety of different-sized and coloured pebbles set in a fine matrix. Some of these pebbles are grey, brown or white, while others are multicoloured.

■ **Composition** Conglomerates are lithified gravels containing particles larger than 2 mm in diameter. Some conglomerates include boulders several metres in diameter. They can contain various pebble types (polymict) or only one pebble type (oligomict). The individual pebbles may be smooth and well rounded from long-term transport and abrasion, or highly angular and jagged, indicating that the sediment was deposited rapidly, with little transport or reworking. If a conglomerate has little matrix then the pebbles tend to be pressed up against each other. These conglomerates are termed clast-supported, because the pebbles (clasts) support each other. If a conglomerate has a high proportion of matrix in which the pebbles are suspended or 'float', then the conglomerate is called a diamictite, i.e. a matrix-supported conglomerate. Diamictite formed by glacial action is termed tillite.

Conglomerate of chert, jasper, quartzite and banded-iron formation pebbles, 12.4 cm. Barberton greenstone belt.

Quartz pebble conglomerate, 22 cm. Witwatersrand goldfield, South Africa.

■ Uses Some conglomerates are used as building stone. Others can contain economic deposits such as gold.

■ Occurrence One of the most famous conglomerates in southern Africa is a glacial tillite, Dwyka tillite, which is a matrix-supported conglomerate found at the base of the Karoo Supergroup sequence of rocks. It is found in **South Africa**, **Namibia**, **Botswana**, **Zimbabwe**, **Mozambique** and **Swaziland**, wherever Karoo-type successions occur. The Dwyka tillite was deposited 300 million years ago by a huge continental ice sheet. Other well-known southern African glacial tillites are the Chuos deposits in Namibia and the Pakhuis tillites in the Western Cape. Famous gold-bearing conglomerates are found in the Archaean Witwatersrand gold mines. Other extensive conglomerate formations are found in the Waterberg, and the Robberg near Plettenberg Bay. Conglomerates often occur in the upper, sedimentary part of ancient greenstone belts.

■ Halite

See also halite in the minerals section.

■ Colour Pure halite is colourless or white. Sometimes halite can contain organic or inorganic impurities and these can produce coloured halite, notably pale pink or pale blue.

■ Composition Halite is a mineral that is classified as a rock when it occurs in sufficient quantities. It is composed of sodium chloride, commonly referred to as rock salt. It is white, soft and can dissolve in water.

■ Uses It is used as a food additive, preservative and flavouring agent.

■ Occurrence Halite forms by evaporation in arid regions and is found in many saltpans in southern Africa. The evaporation causes a concentration of salts in the water, and these usually crystallize around the rim of the deposit. In **Namibia** halite is commercially harvested from saltpans along the coast close to Swakopmund. There are also commercial halite deposits on the east coast of **South Africa** and in dry regions in the central and northwestern parts of the country.

Halite, 5.5 cm. Swakopmund, Namibia.

Limestone, Dolomite

Colour Limestone and dolomite are generally white to grey.

Composition **Limestone** is composed of calcite and/or aragonite, both carbonate minerals, but can contain minor amounts of other carbonate species such as siderite and ankerite. Different types of limestones exist:

- **Chalk** contains the carbonate skeletal remains of micro-organisms.
- **Travertine** is a finely crystalline limestone that has concentric rings or radiating growth patterns. Many stalactites and stalagmites are composed of travertine.
- **Tufa** is a chemical sedimentary rock composed of calcium carbonate that originates around calcium-rich groundwater seepages or springs.
- **Calcrete** comprises calcium carbonate and forms by evaporation that precipitates calcium carbonate in the upper layers of soil.

Note that the carbonate comprising these rocks originates from one of two sources, either organic (i.e. from organisms), as is the case with chalk, or inorganic (i.e. via precipitation of carbonate from a solution).

Dolomite is a calcium-magnesium carbonate mineral species, but dolomite also refers to a rock composed of dolomite crystals. Dolomite can form by evaporation of saline water, but many dolomites originate from the dolomitization of limestone. In this process, magnesium-rich fluids – seawater or percolating groundwater – infiltrate the limestone and chemically replace the limestone with secondary magnesium-rich dolomite.

Uses The main use of limestone is in the manufacture of cement. Both dolomite and limestone are used as fertilizers to neutralize acidic soils. Dolomite is used for refractory bricks and limestone has several other applications, including water and sewage purification, stone dusting in coal mines and as a filler in paint, rubber, wood putty, vinyl tiles and asphalt. Dolomite and limestone are important aquifers, storing vast amounts of underground water and, in some regions, they are notorious for the formation of sinkholes. Limestone and dolomite are common host rocks for important lead and zinc deposits.

Fossil-rich limestone from the Western Cape, South Africa, 10 cm.

Oolitic limestone, 24 cm. Schmidtsdrif, South Africa.

Weathered dolomite, 10.5 cm. Broederstroom, South Africa.

Occurrence Limestone and dolomite deposits are found extensively in southern Africa, e.g. in **South Africa**, in the Northern Cape, west of Vryburg. The world-famous Cango Caves near Oudtshoorn are formed in limestone. Sudwala, Sterkfontein, Makapansgat and Echo Caves are famous dolomite caves. Dolomites and limestones are natural formations for caves because the carbonate minerals easily dissolve in slightly acidic groundwater, leaving vast underground chambers lined with stalactites, stalagmites and other cave deposits. Limestone deposits with travertine and calcrete are found east and northeast of Kimberley and between Johannesburg and Mmabatho. Outcrops of limestone and dolomite are found in all of South Africa's provinces. The dolomite in the Otavi mountainland in **Namibia** is vast and hosts important economic deposits of lead, zinc and vanadium. In **Zimbabwe**, dolomite occurs in the Makonde district, but large commercial deposits are known from the Rushinga dolomite prospect in the Rushinga district. In the Mudzi district dolomite has been quarried together with its host rock, marble, and has been sold for ornamental purposes. The Chinoi Caves are formed in dolomite and limestone.

Mudrocks (includes claystone, shale and siltstone)

Colour Colour is variable, usually grey to black, but also brown, red, purple or mauve.

Composition Mudrocks are the finest grained clastic sedimentary rocks, with individual particles less than a sixteenth of a millimeter in diameter (silt and clay-sized particles). Mudrocks originate when clay and silt particles settle underwater, or are deposited on land by the wind. Mudrocks are usually fairly soft rocks and erode easily, forming gentle slopes or flat valleys. If a mudrock is made up exclusively of clay-sized particles (less than 3.9 microns in diameter), it is called a claystone. If it is made up of only silt-sized particles (3.9–62.5 microns), it is called a siltstone. Usually, silt and clay-sized particles occur together, hence the general name of 'mudrock' for these

Layers of grey mudrock with lighter sandstone beds exposed in a cliff face. Newcastle, KwaZulu-Natal.

fine-grained sedimentary rocks. A finely layered mudstone is called shale. Loess is a special category composed of windblown silt-sized particles.

■ **Occurrence** Mudrocks are common throughout many sedimentary geological formations in southern Africa. In **South Africa** the topography of large parts of the Karoo region, the Western Cape, Eastern Cape, Free State and KwaZulu-Natal is the result of differential weathering of mixed sandstone and mudrock. The Springbok Flats are produced by weathered mudrock and deposits of the same kind occur in parts of the Waterberg. In **Namibia**, similar 'Karoo' mudrocks outcrop in the Toscanini and Waterberg regions. Mudrock/shale is found interbedded with sandstones in the Nama Group of rocks in southern Namibia. **Zimbabwe** has extensive mudrock sequences. Mudrock is associated with coal in the Wankie coalfield and in the southeast of the country. Mudrock also outcrops in large parts of eastern **Swaziland**.

Finely laminated shale, 12 cm. Northern KwaZulu-Natal, South Africa.

Pyroclastic rocks

Colour Variable from grey to brown.

Composition Pyroclastic sediments consist of volcanic material transported and deposited by wind, water or gravity and hence they are sedimentary. When volcanoes erupt subaerially (on or just above the Earth's surface), ash and volcanic debris are blasted into the atmosphere. Dust-sized ash particles can be swept vast distances by the wind before settling to the ground, either on dry land or in waterbodies such as lakes or oceans. Once deposited, the particles, being unstable under normal pressures and temperatures, rapidly undergo chemical changes and are generally transformed into clay minerals. Volcanic tuff alters to clay, volcanic glass devitrifies and larger particles disintegrate. It is not always easy to recognize ancient pyroclastic deposits because they seldom resemble their original composition. The three types of pyroclastic rock are agglomerate (a conglomerate made up of large pyroclastic fragments), tuffs (consolidated fine-grained volcanic ash) and ignimbrite (very fine tuff material that is welded together).

Classification of pyroclastic rocks		
Particle size (mm)	Description of sediment	Rock type
>64 mm	block, bomb	agglomerate; pyroclastic breccia
2–64 mm	lapilli	lapilli tuff
<2 mm	ash	tuff/ignimbrite

Occurrence Pyroclastic rocks are relatively rare in the ancient rock record because they are chemically unstable and weather readily, breaking down into clay minerals. Some occur in the Loskop Formation in the Bushveld Complex, **South Africa**. They also occur in layers between the lavas of the Ventersdorp Supergroup, the Soutpansberg Group, the Rooiberg Group of the Bushveld Complex, the Pilanesberg and other alkaline complexes, and in some of the ancient greenstone belts. Pyroclastic phases are present in some well-preserved kimberlite pipes. There are thin pyroclastic layers in the lower Karoo Supergroup sedimentary rocks in the southern Cape region.

Tuffaceous volcaniclastic rock, 7 cm.
Du Toitspan kimberlite, South Africa.

Sandstone (incorporating arkose, greywacke, and quartz- and lithic arenite)

Colour Commonly white, grey or cream-beige, but can be red to orange depending on the mineral composition and clay matrix content.

Composition Sandstones are made up of sand-sized particles that range in size from 62.5 microns to 2 mm in diameter. Quartz is the most common mineral component in sandstones, followed by feldspars and clay minerals. Other accessory minerals in sandstone depend on the original source rock that was weathered to form the sandy sediment. Some varieties of sandstone are named according to their composition. The following are important varieties:

- **Arkose** has a large proportion of quartz and alkali feldspar, but not much clay.
- **Greywacke** has a high clay content and includes other minerals and small rock fragments.
- **Quartz arenite** is made up almost entirely of quartz grains with very few other minerals.
- **Lithic arenite** is composed of a mixture of quartz and small rock fragments.

Because sandstones were transported and deposited by water or wind, they commonly display sedimentary structures such as ripple marks and cross-bedding. The different types of structures usually reflect prevailing hydrodynamic/aerodynamic conditions during sand transport and deposition.

Uses Certain sandstones have been used as building material, e.g. at the Union Buildings in Pretoria. Many historical houses, churches and municipal buildings in the Free State, Western Cape, Eastern Cape

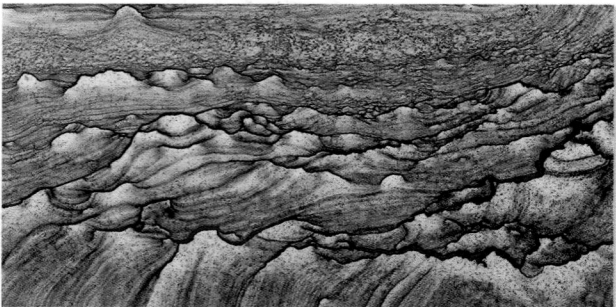

'Picture' sandstone, 14.5 cm. The intricate patterns are produced by iron staining. Namibia.

Large cross-bedding in sand dunes north of Richards Bay, South Africa.

and southern Mpumalanga are built from local sandstone. More recently, sandstone has been used for decorative tiles and table tops.

Occurrence Sandstones are typically hard rocks. Resistant to erosion, they form prominent and spectacular cliffs and mountains. One of the most famous outcrops of sandstone in southern Africa is Table Mountain, overlooking Cape Town. The distinctive layering of these sandstones can clearly be seen. The foothills of the Drakensberg mountain range in **South Africa** and **Lesotho** are composed predominantly of sandstone, and the Golden Gate National Park in the eastern Free State contains spectacular outcrops of the Clarens Formation sandstones. The Blyde River Canyon in South Africa and the equally scenic Fish River Canyon consist of layers of sandstone with some mudstone, sculptured by millions of years of erosion. These rocks form part of the Nama sedimentary sequence that outcrops over large portions of southern **Namibia**. Karoo Supergroup rocks contain thick sandstone deposits and these occur in all of the southern African countries including **Botswana**, **Zimbabwe**, **Swaziland** and **Mozambique**.

Red, orthoclase-rich arkose sandstone, 7 cm. Waterberg region, South Africa.

Metamorphic rocks

Rocks subjected to intense heat, pressure and chemical reaction over long time periods are transformed into metamorphic rocks. This is because these geological and chemical processes cause the minerals to recrystallize. Considerable pressure often imparts a foliation or alignment of minerals in the newly formed metamorphic rock. Schist and gneiss are examples of rocks with layered or foliated minerals.

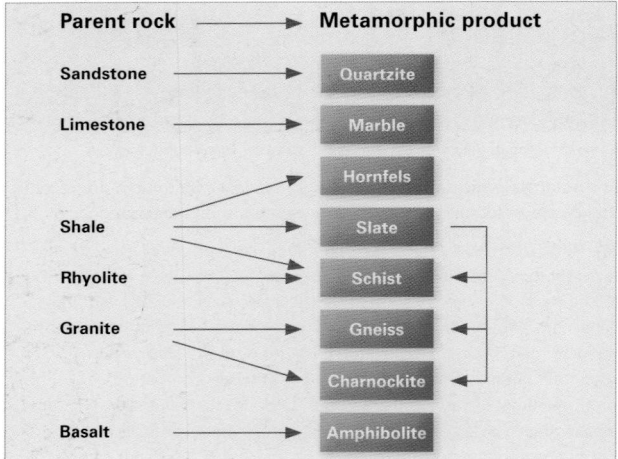

Parent rock		Metamorphic product
Sandstone	→	Quartzite
Limestone	→	Marble
		Hornfels
Shale	→	Slate
Rhyolite	→	Schist
Granite	→	Gneiss
		Charnockite
Basalt	→	Amphibolite

The original composition of the host rock will determine what type of metamorphic minerals or rock will form. Basalt will produce a dark, coarse amphibolite composed of iron-rich silicate minerals. A limestone, composed of calcite, forms a light-coloured marble. When sandstone is subjected to metamorphism it forms a quartzite, which is still composed of quartz grains, but these often recrystallize and are welded together by silica cement.

Except for hornfels, all the minerals listed in the figure above are produced by *regional metamorphism*, i.e. the rocks have been subjected to widespread heat and pressure deep in the Earth's crust. A second form of metamorphism, *contact metamorphism*, occurs when rocks are exposed to the hot surface along the contact zone of an igneous intrusion. For instance, hot igneous intrusions of dolerite that intersect sandstones and shales may 'bake' these sedimentary rocks along their contact margins. The metamorphic rock that is produced by contact metamorphism is called a hornfels.

The intensity of metamorphism that affects a parent rock is referred to as the metamorphic grade. Three levels of intensity are recognized:

Occurrence In **South Africa**, Marble Delta in KwaZulu-Natal covers an area of approximately 40 km² inland of Port Shepstone, close to Oribi Gorge. The marbles occurring here have been repeatedly folded and intruded by several granites and charnockites, and they contain an array of metamorphic minerals. These marbles are periodically worked and quarried. Marble is also found in the Marble Hall district, Mpumalanga, and in the Vanrhynsdorp district in the Western Cape. In **Namibia**, important economic deposits of marble are found in the Karibib and Usakos districts and several are commercially quarried. Marbles are found in eastern **Botswana**, north of Baines Drift. In **Zimbabwe**, marble is exploited from the Rogo limestone near Nyamapanda. Attractive banded marble has also been quarried in the Hurungwe area. Several Zimbabwean geological sequences contain marble: the Rushinga metamorphic suite, and the Lomagundi, Beitbridge and Umkondo groups.

Quartzite

Colour White, off-white, cream or grey. Quartzite can also be glassy.

Uses Occasionally used as dimension stone.

Composition Quartzite is a metamorphosed sandstone composed of recrystallized quartz and very few other minerals. The original grains become cemented and 'welded' together by silica solution that forms during metamorphism. 'Glitterstone' is a common name for quartzite with crystals of muscovite (white mica) in its matrix.

Occurrence Quartzite is found in many areas of southern Africa. In **South Africa**, it occurs in the Witwatersrand goldfields and there are large outcrops in northern KwaZulu-Natal. Quartzites outcrop in **Namibia**, in the Windhoek district, and some of these have been used for building purposes. Quartzites are fairly abundant in the Nama sedimentary succession. They are also present in the Chobe district of **Botswana**. Quartzite occurs in **Zimbabwe** near Kariba, in the Makuti Group. It was also exploited from the northeastern parts of the country and is used as a paving stone. Zimbabwe has a green quartzite referred to as 'emerald slate'. The green colour in the metamorphosed sandstone is caused by crystals of fuchsite, a chrome-bearing mica. It is quarried for floor tiles from a few deposits located north and west of the Sandawana emerald mine in the Mberengwa District.

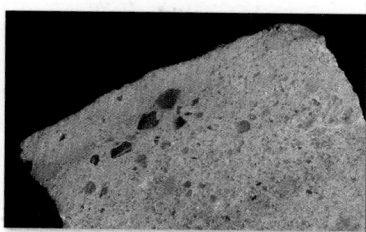

Quartzite containing small grey chert pebbles, 7.8 cm. Northern KwaZulu-Natal, South Africa.

Schist

Colour Variable, but most schist has a shimmer that is caused by the presence of abundant mica, usually muscovite.

Composition Schist is an intermediate-grade metamorphic rock in which the minerals are aligned parallel to one another, giving it a texture known as 'schistose'. Micas, usually biotite or muscovite, are the most common minerals found in schists, but amphiboles, e.g. hornblende, can also form schists. New metamorphic minerals, such as andalusite, garnet and staurolite, form large crystals in some schists.

Occurrence In **South Africa**, schist is common in the Archaean greenstone belts in Mpumalanga and Limpopo and also in parts of the Northern Cape. There are large and continuous outcrops of schist in central **Namibia** in the Khomas Hochland, and in the Kuiseb Valley and the area around the Gamsberg. In **Zimbabwe**, schist is common in the many greenstone belts and in the Zambezi Valley. Southern African greenstone belts have extensive schist formations – commonly chlorite and/or talc schist.

Biotite schist, 16 cm. Murchison greenstone belt, South Africa.

Serpentinite

Colour Serpentinite is a soft grey to dark green to grey-black metamorphic rock.

Composition A rock composed of serpentine minerals is called a serpentinite. Serpentine is actually the common name used for a group of three different minerals – lizardite, antigorite and clinochrysotile. Serpentine minerals are found in metamorphic rocks with a high magnesium content. Talc and actinolite are informally included with the serpentine minerals.

Uses Some lapidary varieties of serpentinite are 'bowenite', a translucent green and/or yellow-green variety and 'soapstone', a massive grey-green rock. (The term 'soapstone' is often used to describe any soft, coloured rock that can be carved, so may include talc-rich rocks that are not serpentinite.)

Occurrence In **South Africa**, serpentinites are found in the greenstone belts of Mpumalanga. In **Namibia**, grey-green soapstone serpentinite called 'steatite' is found on the farm Omieve 179, 120 km northeast of Windhoek, where it originates from altered ultramafic rocks. Serpentinites are also found in intrusive rocks approximately 50 km northeast of Seeis, and southeast of Windhoek. Christmas Kop in **Botswana**, 1 km north/northwest of Francistown, contains serpentinite. It outcrops in northeast Botswana, 6 km north of

Green and grey serpentinite, 8.5 cm. Kaapsehoop district, South Africa.

Francistown, near the Ntshe River and at the confluence of the Shashe and Tati rivers at Pawmaitse. In **Zimbabwe** soapstone occurs in the Nyanga area and a beautiful translucent pale green variety comes from the Selukwe chrome mines. The famous Shona sculptors frequently use soapstone as their carving material. 'Bowenite' is found in the Mashava area. Talc serpentinites are found in the Mutare, Masvingo, Mberengwa and Kwekwe districts, and the Sinoia and Gweru areas. At Gweru, an attractive green serpentinite occurs that has veins of quartz and calcite and minute specks of golden coloured chalcopyrite. An unusual and attractive translucent pale green variety is known from the chrome mines at Shurugwi. Serpentinite also outcrops in northwest **Swaziland**.

■ Slate

■ **Colour**   Grey to black.

■ **Composition**   Slate is a relatively low-grade metamorphic rock, formed by the metamorphism of shale. It has very good cleavage, but the constituent minerals have not recrystallized, so slate is very fine-grained. When new metamorphic minerals do begin to crystallize in slates, they form large crystals, such as andalusite.

■ **Uses**   Used for paving stones, floor tiles and roof tiles.

■ **Occurrence**   Slate is most commonly found close to the contact zones of large igneous intrusions. Several slate quarries are located in the North West province in **South Africa**, where the Bushveld Complex intruded into the shales of the Transvaal Supergroup.

Slate, 9.2 cm. Zeerust district, South Africa.

GLOSSARY

Adamantine – describing a very high lustre, like that of diamond.

Aggregate – a mass, or assemblage, of minerals.

Alkaline – having the properties of an alkali substance, e.g. one that contains potassium or sodium.

Alluvial – relating to alluvium, i.e. detrital material (clay, sand, gravel) transported and deposited by a river.

Alpine cleft deposit – cracks or crevices that developed in high mountain ranges and which contain crystals of quartz, adularia and associated minerals.

Amorphous – describing minerals that lack crystalline structure and have no characteristic external form.

Amphibole(s) – a group of dark coloured, iron-magnesium-rich silicate minerals.

Amphibolites – metamorphic rock consisting mainly of amphiboles and lesser amounts of plagioclase.

Amygdale – a gas cavity (vesicle) in certain extrusive igneous rocks, filled with secondary minerals such as quartz, calcite, zeolites or chalcedony.

Archaean – relating to the geological eon in which the earliest known rocks were formed.

Assemblage – a sequence or aggregate of minerals.

Bedded – occurring in layers.

Botryoidal – resembling a bunch of grapes; a mineral having spherical shapes on its surface.

Bowtie crystals – crystals that are grouped together in the shape of a bowtie.

Breccia/brecciation – a coarse-grained rock composed of angular fragments and/or grains; the forming of breccia.

Calc-silicate – comprised mainly of calcium and silicon.

Carat – unit of weight of diamonds, gold or pearls; 1 carat = 0.2 grams.

Carbonate mineral – a mineral containing the anionic structure CO_3^2 e.g. calcite, $CaCO_3$.

Carbonatite – a carbonate rock, usually of magmatic origin, and commonly associated with alkaline rocks.

Cauliform – the rounded, globular texture of some minerals.

Chert – an amorphous sedimentary rock type that is composed of silica.

Chromophore – a molecule that imparts colour in a mineral.

Clastic – comprised of fragments of older rocks and minerals.

Clay – fine sedimentary grains less than 4 microns in diameter.

Conchoidal – having a fracture pattern characterized by hollowed or rounded surfaces, i.e. the mineral does not break along fracture planes.

Contact metamorphism – the process whereby changes occur in the chemistry and physical structure of rocks directly in contact with intruding magma.

Craton – a stable, usually ancient, part of the Earth's continental crust that has undergone little structural deformation.

Crypto-crystalline – describing crystals so fine that they are only visible under high magnification.

Crystal – an homogeneous object composed of a chemical element, compound or mixture, having a regular, repeated arrangement of atoms that can be outwardly expressed by plane faces.

Dendrite – a structure with a tree-like form.

Deposit – an economic orebody, or an accumulation of sediments.

Detritus – loose rock and mineral matter produced by mechanical weathering.

Dewatering – the process whereby water is removed from a wet solid.

Diagenesis – a chemical, physical or biological change that transforms unconsolidated material into rock.

Dipyramidal crystals – crystals formed of two symmetrical pyramids base to base.

Dolerite – synonymous with diabase: an intrusive, basic igneous rock composed of pyroxene and plagioclase feldspar, and minor olivine.

Druse/drusy – a crusting of small, well-formed crystals, e.g. in a geode.

Dyke – a tabular, inclined igneous intrusion that cuts through and across surrounding rocks.

Ejecta – matter ejected from a volcano.

Epimorph – The growth or deposition of a layer of one mineral over another. The underlying crystal may dissolve away leaving a 'cast' or mould showing the original shape of the dissolved crystal.

Euhedral – describing well-formed crystals with recognisable faces.

Extrusion/extrusive – the outpouring of magma as lava.

Fishtail crystals – twinned crystals that are V-shaped.

Foliated – the texture or structure in a rock characterized by a planar arrangement of minerals.

Gabbro – a dark, mafic igneous rock comprising mainly plagioclase, pyroxene and some olivine.

Gemstone – any mineral or rock that can be cut and/or polished,

is beautiful, durable and rare, and that can be used for jewellery or ornamental purposes.

Geode – a hollow, subrounded or elongate cavity lined with drusy crystals. Occurs in limestones and lavas.

Gneiss – a foliated, generally coarse-grained rock formed by regional high-grade metamorphism.

Gossan – a weathered, residual rock deposit capping a sulphide deposit, from which minerals have been leached, leaving behind reddish iron oxide.

Greenstone belt – generally, small isolated portions of the Earth's ancient crust containing highly deformed rocks, often coloured green by secondary minerals, e.g. chlorite, fuchsite, actinolite and epidote.

Greisen – a granitic rock type composed of quartz, mica, topaz and tourmaline. Greisen can also contain fluorite, cassiterite and ferberite.

Greywacke – a variety of sandstone composed of angular grains of quartz and feldspar in a clay matrix.

Heavy-mineral sands – sands where grains of minerals with a high specific gravity are concentrated.

Hornfels – a fine-grained metamorphic rock formed by contact metamorphism.

Hydrated/hydration – describing a mineral in which water (H_2O) is part of the chemical composition.

Hydrothermal – of or relating to heated water, from which minerals or a mineral deposit may be precipitated.

Igneous rock – a rock that crystallized from molten magma.

Inclusion/included – material that became trapped inside a mineral when it formed.

Intrusion/intrusive – a body of magma that has cooled and crystallized below the surface of the Earth.

Kimberlite – a volcanic rock containing olivine, other magnesium- and iron-bearing minerals and certain micas. A host rock for diamonds.

Lapidary – relating to the cutting and polishing of stone for decorative use.

Lineation – lines or bands in a rock caused by the alignment of minerals.

Lithification/lithify – the process whereby sediments are consolidated to form rock.

Mafic – describing a group of dark-coloured igneous rocks, rich in ferromagnesium-bearing minerals.

Metamorphic rock – rock that has been transformed into a new rock type by heat and pressure.

Metamorphism – the recrystallization of rock that results from subjection to high heat and pressure.

Microcrystals – crystals invisible to the naked eye.

Norite – a coarse-grained igneous rock containing plagioclase (labradorite) and orthopyroxene (hypersthene).

Opencast – describes mining operations where the ore is removed from an open excavation.

Outcrop – a geological formation or rock that is exposed on the Earth's surface.

Oxidation zone – the part of a sulphide orebody that has been exposed at, or close to, the Earth's surface and that has oxidized.

Pegmatite – a very coarse-grained igneous rock, occurring in lenses or veins. Pegmatites are the last phase to crystallize from a magma and often contain high concentrations of trace elements as well as large quantities of rare elements and minerals.

Pipe – a cylindrical orebody. Kimberlite, for instance, often forms in pipes.

Placer – describes an economic deposit formed when mechanical weathering of bedrock by water or by glacial action concentrates mineral particles.

Pluton – an igneous intrusion, usually deep-seated and coarse-grained.

Polymetallic – composed of different metals.

Polymorphism – the crystallization of a chemical substance in more than one form.

Pseudocubic – resembling a cubic crystal, but not belonging to the cubic crystal system.

Pseudomorph – a mineral whose outward crystal form resembles that of a different mineral species.

Pyroclastic – a clastic sedimentary rock composed of ejecta emitted from a volcano.

Pyroxene – a group of minerals containing iron and magnesium that occur in mafic and ultramafic igneous and metamorphic rocks.

Quartzite – a metamorphosed sandstone composed predominantly of quartz grains.

Rare-earth elements – a group of 17 elements running consecutively from lanthanum to lutetium in the periodic table.

Refractory – describing an ore that does not readily yield its valuable constituents, or a mineral that can withstand high temperatures.

Reniform – resembling a kidney.

Rhyolite – the extrusive equivalent of a granite; a lava composed of quartz, orthoclase feldspar and other alkali feldspars.

Schist/schistosity – a layered metamorphic rock that is commonly composed of mica minerals, amphiboles or talc.

Secondary vein filling – a vein filled by minerals and/or crystals that are secondary in origin, i.e. they formed after the host rock formed.

Sedimentary rocks – rocks formed from detrital, fragmental material transported by wind, water or ice and then deposited. Also refers to rocks formed from the chemical sediments that are precipitated from a solution or crystallize due to evaporation, and from organic sediments, such as coal.

Serpentinite/serpentinize – rocks in which the original minerals have converted into the mineral serpentine.

Silicate mineral/siliceous – a mineral whose crystal structure contains SiO_4 tetrahedra.

Skarn – metamorphic rock that is composed mainly of calcium silicate minerals, usually pyroxene and garnet, but that may also contain grossular, diopside, olivine, scapolite, epidote, vesuvianite, axinite, amphiboles and prehnite.

Stellate – star-shaped or radiating.

Subadamantine – a lustre that is slightly duller than adamantine.

Submetallic – a lustre similar to that of metal, but not quite as bright.

Subrounded – not quite rounded in shape.

Tailings – material remaining after minerals are recovered from an ore.

Terrain – a general term used to describe a group of rocks and the area in which they outcrop.

Tuff – a volcanic rock composed of compacted volcanic ash and volcanic dust.

Twin – an ordered intergrowth of two or more single crystals of the same mineral.

Vug – a small cavity in a vein or rock, usually lined with crystals.

Xenolith – rock fragments that are foreign to the body of igneous rock in which they occur; an inclusion.

Zeolite – a kind of silicate mineral.

CREDITS & ACKNOWLEDGEMENTS

All photographs were taken by the author. All specimens belong to the author, except for the following: Council for Geoscience: 64 (top); CS Diamonds: 12, 47 (centre); De Beers: 47 (bottom); Desmond Sacco: 8 (top), 24 (bottom left), 57 (bottom), 113 (centre); Eric Farqharson: 36; Johannesburg Geological Museum: 27; Private collection: 46, 47, 60, 61; Rob Smith: 94 (bottom left); Uli Bahmann: 7 (top), 95 (top), 102 (bottom left), 103 (top); University of Johannesburg Geology Department: 126, 127, 128 (top), 129, 132, 141, 143, 151, 152.

The collectors and institutions listed above are thanked for loaning specimens for photography. I would also like to thank Pippa Parker of Random House Struik for suggesting the concept of this book. Emily Bowles and Louise Topping edited and designed the book and I am very grateful for their skills in this regard.

BIBLIOGRAPHY

Anhaeusser, CR and Maske, SH (Eds).
 1986. *Mineral Deposits of southern
 Africa*, Vols I & II. Geological Society
 of South Africa: Johannesburg,
 2 335 pages.

Baldock, JW. 1977. *Resources
 Inventory of Botswana: Metallic
 Minerals, Mineral Fuels and
 Diamonds*. Mineral resources
 report no. 4, Geological Survey
 Department: Lobatse, 69 pages.

Cairncross, B. 2000. *The Desmond
 Sacco Collection: Focus on Southern
 Africa*. D. Sacco: Johannesburg,
 408 pages.

Cairncross, B. 2004. *Field Guide to
 Rocks and Minerals of Southern
 Africa*. Random House Struik:
 Cape Town, 292 pages.

Cairncross, B and Bahmann, U. 2006.
 Famous Mineral Localities, Erongo,
 Namibia. *Mineralogical Record 37*:
 361–470.

Cairncross, B, Beukes, NJ and Gutzmer,
 J. 1997. *The Manganese Adventure.*
 Associated Ore & Metal Corporation:
 Johannesburg, 250 pages.

Cairncross, B and Dixon, R. 1995.
 Minerals of South Africa.
 Geological Society of South Africa:
 Johannesburg, 289 pages.

Carney, JN, Aldiss, DT and Lock, NP.
 1994. The Geology of Botswana.
 Mineral Department Bulletin 0037,
 Geological Survey Department:
 Gaborone, 113 pages.

Daltry, VDC. 1992. Type mineralogy
 of Namibia. Bulletin 1, Directorate
 Geological Survey, Ministry of Mines
 and Energy: Windhoek, 142 pages.

Hunter, DR. 1961 (1991). The Geology
 of Swaziland. Swaziland Geological
 Survey and Mines Department:
 Mbabane, 104 pages.

Jahn, S, Medenbach, O, Niedermayr,
 G and Schneider, G. 2001. *Namibia
 zauberwelt edler steineund kristalle*.
 Bode Verlag GmbH: Haltern,
 224 pages.

Johnson, MR, Anhaeusser, CR and
 Thomas, RJ. (Eds). 2006. *The
 Geology of South Africa*. Geological
 Society of Africa: Johannesburg/
 Council for Geoscience, Pretoria,
 691 pages.

Lachelt, S. 2004. *Geology and Mineral
 Resources of Mozambique*. Council
 for Geoscience: Pretoria, 515 pages.

Macintosh, EK. 1988. *Rocks, Minerals
 and Gemstones of Southern
 Africa: A Collector's Guide*. Struik
 Publishers: Cape Town, 120 pages.

McIver, JR. 1966. *Gems, Minerals and
 Rocks in southern Africa*. Purnell and
 Sons: Johannesburg, 267 pages.

Von Bezing, L. 2007. *Namibia Minerals
 and Localities*. Bode Verlag GmbH:
 Haltern, 856 pages.

Warner, SM. 1972. *Check List of the
 Minerals of Rhodesia*. Rhodesia
 Geological Survey, Bulletin No. 69:
 Salisbury, 101 pages.

INDEX

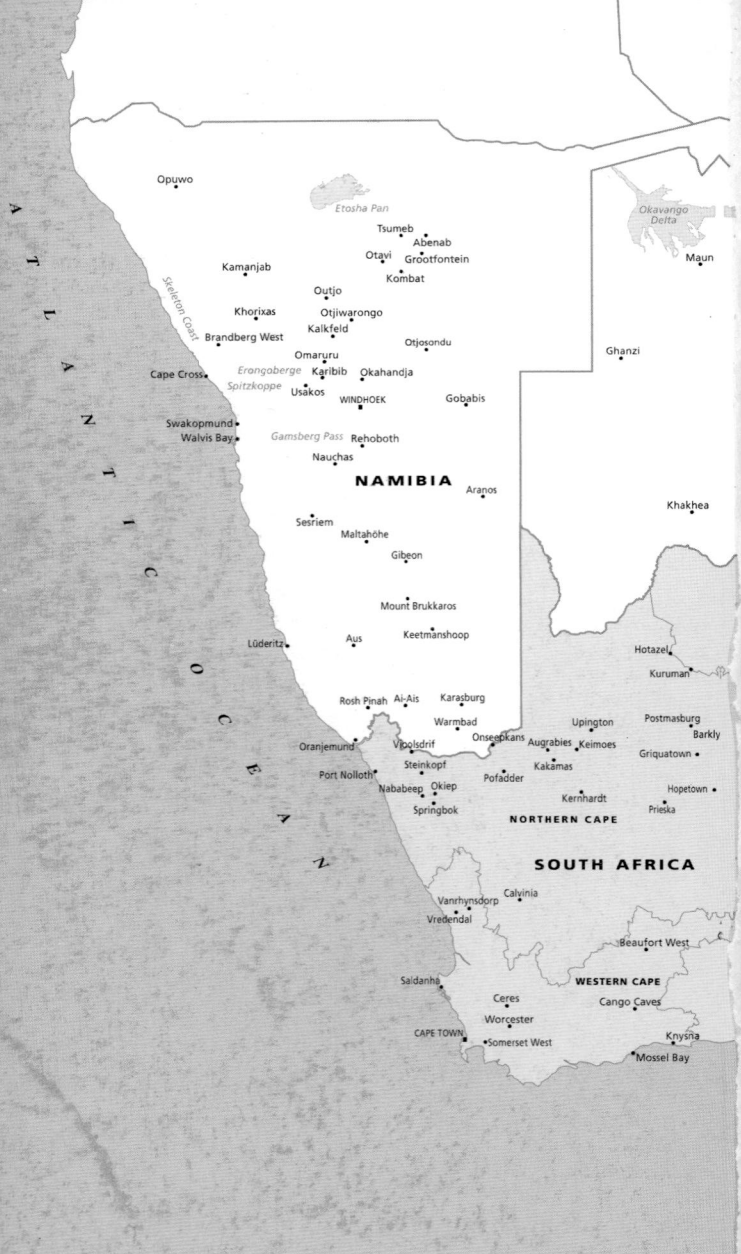